THEORETICAL CONCEPTS AND
HYPOTHETICO-INDUCTIVE INFERENCE

SYNTHESE LIBRARY

MONOGRAPHS ON EPISTEMOLOGY,

LOGIC, METHODOLOGY, PHILOSOPHY OF SCIENCE,

SOCIOLOGY OF SCIENCE AND OF KNOWLEDGE,

AND ON THE MATHEMATICAL METHODS OF

SOCIAL AND BEHAVIORAL SCIENCES

Editors:

DONALD DAVIDSON, *The Rockefeller University and Princeton University*

JAAKKO HINTIKKA, *Academy of Finland and Stanford University*

GABRIËL NUCHELMANS, *University of Leyden*

WESLEY C. SALMON, *Indiana University*

THEORETICAL CONCEPTS AND HYPOTHETICO-INDUCTIVE INFERENCE

by

ILKKA NIINILUOTO

Humanities Research Council, Academy of Finland

and

RAIMO TUOMELA

Department of Philosophy, University of Helsinki

D. REIDEL PUBLISHING COMPANY

DORDRECHT-HOLLAND / BOSTON-U.S.A.

Library of Congress Catalog Card Number 73–83567

ISBN 90 277 0343 4

Published by D. Reidel Publishing Company,
P.O. Box 17, Dordrecht, Holland

Sold and distributed in the U.S.A., Canada and Mexico
by D. Reidel Publishing Company, Inc.
306 Dartmouth Street, Boston,
Mass. 02116, U.S.A.

To Jaakko Hintikka

CONTENTS

PREFACE

Conceptual change and its connection to the development of new scientific theories has recently become an intensively discussed topic in philosophical literature. Even if the inductive aspects related to conceptual change have already been discussed to some extent, there has so far existed no systematic treatment of inductive change due to conceptual enrichment. This is what we attempt to accomplish in this work, although most of our technical results are restricted to the framework of monadic languages. We extend Hintikka's system of inductive logic to apply to situations in which new concepts are introduced to the original language. By interpreting them as theoretical concepts, it is possible to discuss a number of currently debated philosophical and methodological problems which have previously escaped systematic and exact treatment. For instance, the role which scientific theories employing theoretical concepts may play within inductive inference can be studied within this framework. From the viewpoint of scientific realism, such a study gives outlines for a theory of what we call hypothetico-inductive inference.

Some parts of this work which are based on Hintikka's system of inductive logic are fairly technical. However, no previous knowledge of this system is required, but, in general, acquaintance with the basic ideas of elementary logic and probability theory is sufficient.

This work is part of a project, originated by Professors Jaakko Hintikka and Raimo Tuomela, concerning the role of theoretical concepts in science. Within this project, the study of the inductive aspects of this role was started in 1970 by Ilkka Niiniluoto. Works by Niiniluoto in which this study has been reported include the following: 'Can We Accept Lehrer's Inductive Rule?', *Ajatus* **33** (1971) 254–265; 'Inductive Systematization: Definition and a Critical Survey', *Synthese* **25** (1972) 25–81; 'Inductive Systematization and Empirically Trivial Theories', in *Logic, Language and Probability, A Selection of Papers Contributed to Sections 4, 6, and 11 of the IVth International Congress of Logic, Methodology and Philosophy of Science, Bucharest*, 1971 (ed. by R. Bogdan and I. Niiniluoto),

D. Reidel, 1973; and jointly with Jaakko Hintikka, 'On Theoretical Terms and Their Ramsey-Elimination: An Essay in the Logic of Science' (in Russian), *Filosofskiye Nauki* **1** (1973) 49–61. Tuomela has discussed theoretical concepts within induction in 'The Role of Theoretical Concepts in Neobehavioristic Theories', *Reports from the Institute of Philosophy, University of Helsinki*, No. 1, 1971; and in *Theoretical Concepts*, Library of Exact Philosophy, vol. 10, Springer-Verlag, 1973. The latter is a systematic study of theoretical concepts and their role within scientific theorizing.

The study which led to this monograph was started in 1971 by the authors. This work, especially its philosophical and methodological content, was developed by means of intense discussions and detailed mutual criticism between the authors. As this work is a result of a joint attempt, in many cases it is not easy to distinguish the individual contributions of the authors. This applies at least to Chapters 1.2, 8.1, 8.3, and 11.5. Of the remaining part, Niiniluoto can be regarded as primarily responsible for Chapters 2–6, Chapter 7.4, Chapter 9, Chapter 10.1, and Chapter 11.1–4, and Tuomela for the rest.

We dedicate this work to Professor Jaakko Hintikka. This is not only a token of our deep gratitude for the encouragement we have received from him in completing this project, and in our philosophical work in general. In writing this monograph, Hintikka's studies in inductive logic and in semantic information theory have formed an indispensable starting point for us. An attempt to extend Hintikka's inductive system to cover a so far unexplored field is the best tribute we can pay to his fertile ideas on induction.

During the writing of the present work, Niiniluoto has held a research assistantship at the Humanities Research Council (Valtion humanistinen toimikunta) in the Academy of Finland. Much of Tuomela's work was done in Montreal while he was on leave from his professorship at the University of Helsinki, and held a Killam postdoctoral fellowship granted by Canada Council.

This study has partly been supported by the Emil Aaltonen Foundation (Emil Aaltosen säätiö), which is gratefully acknowledged. We wish to thank Mr. Fred A. Fewster for checking the language of the present work.

ILKKA NIINILUOTO

Helsinki, December 1972   RAIMO TUOMELA

CHAPTER 1

THEORETICAL CONCEPTS AND INDUCTIVE INFERENCE

This work represents an attempt to clarify the role which theoretical concepts may play within inductive scientific inference or inductive systematization. A study is made of some of the possible gains that accrue from the introduction of theories employing new theoretical concepts. We are especially interested in gains for which theoretical concepts are logically indispensable, as they provide us with strong methodological reasons for the introduction and employment of theoretical concepts. We have restricted ourselves to dealing with relatively simple kinds of theories, which may be taken to exemplify theories from the primarily non-quantitative sciences.

The role of theoretical terms within hypothetico-deductive theorizing has recently been discussed rather intensively. However, so far, no systematic account has existed of their function within inductive inference. Even the very possibility of such an account has been questioned. In contrast with this, our work is an attempt to develop a framework for discussion of the role of theoretical concepts within what will be termed hypothetico-inductive inference. Such a framework has to cope with a number of problems not present in deductive inference. Perhaps the best known of these problems are those resulting from the lack of reasonable transitivity properties of inductive inference. As an introduction to the problems discussed in this study, we begin by showing how a 'transitivity dilemma' arises in connection with theoretical concepts and inductive inference. Another related problem discussed in this introductory chapter is how one can account for the effect of conceptual change on induction.

1. PROBLEMS OF INDUCTIVE SYSTEMATIZATION: THE TRANSITIVITY DILEMMA

1.1.   In his well-known article 'The Theoretician's Dilemma', of 1958, Carl G. Hempel argued that even if theoretical concepts may be logically dispensable for the deductive systematization of observational statements, they are not dispensable for inductive systematization. However, Hempel did not give a general account of inductive systematization in his paper, but argued only by means of an example. It has been claimed that his example does not show what it purports to show, and thus new arguments are needed. Recently, the indispensability of theoretical concepts has been discussed by a number of authors (cf. Niiniluoto (1973a) for a critical review of them). In this work, we make an attempt to discuss the logical indispensability of theoretical concepts for inductive systematization within a setting that differs from that of the other authors. The problem will be discussed by means of an applied form of Hintikka's inductive logic. Furthermore, our discussion is not restricted to the question of logical indispensability, but will also concern the general role and the methodological desirability of theoretical concepts within some interesting cases of inductive systematization, such as inductive explanation.

Let us start by discussing one of Hempel's examples of inductive systematization. We consider a simple theory T formalized within monadic predicate logic. As observational predicates, we have 'soluble in ether' ('E'), 'bursts into flame' ('F'), 'has a garlic-like odor' ('G'), 'produces skin burns' ('S'), 'soluble in turpentine' ('T'), 'soluble in vegetable oils' ('V'). As theoretical predicates, we have 'white phosphorus' ('P') and 'has an ignition temperature of 30°C' ('I'). As axioms of T, we have first the following 'correspondence rule' which gives five observational necessary conditions for P:

(1.1) $(x) [P(x) \supset (E(x) \, \& \, G(x) \, \& \, S(x) \, \& \, T(x) \, \& \, V(x))]$.

Next we have the following 'purely theoretical' statement:

(1.2) $(x) (P(x) \supset I(x))$.

Finally we have the 'correspondence rule'

(1.3) $(x) (I(x) \supset F(x))$.

Hempel's argument now goes as follows. Let a be an individual showing the symptoms of P, that is, $E(a) \, \& \, G(a) \, \& \, S(a) \, \& \, T(a) \, \& \, V(a)$. This evidence is supposed to be an inductive sign of $P(a)$, or to give inductive support to $P(a)$. In other words, it is supposed that we can inductively infer $P(a)$ on the basis of (1.1) and our evidence. Next, we deductively infer $I(a)$ on the basis of (1.2) and $P(a)$. Finally, we deduce the new observational consequence $F(a)$ from $I(a)$ and (1.3). Of course, the important thing here is that the conclusion $F(a)$ is not deductively obtainable from the theory T and the evidence, but that an inductive leap is needed in the inference. As this inductive step involves the theoretical concept P, the argument takes this to show the indispensability of theoretical concepts for the inductive systematization of observational statements.

It can be and it has been argued that the above example as well as other similar examples by Hempel and Scheffler (cf. Hempel, 1963 and Scheffler, 1963) do not show what they purport to show. There are two main types of reasons given in the counterarguments (cf. the survey in Niiniluoto, 1973a). First, it has been argued that, in examples of the above kind, theoretical concepts do not establish inductive systematization. Secondly, it has been argued that whenever a theory achieves inductive systematization, the same inductive systematization can be achieved by means of a suitably constructed observational subtheory of the original theory.

For discussion of the first kind of counterargument we shall make the assumption that inductive inference is essentially connected with the logical form of statements. Let us then somewhat simplify the theory, to have a new theory T, which is axiomatized by the conjunction of the following two statements:

(1.4) $(x)\,(P(x) \supset O_1(x))$
(1.5) $(x)\,(P(x) \supset O_2(x))$.

Here 'P' is a theoretical predicate as before and 'O_1' and 'O_2' are observational predicates (cf. below for a clarification of the theoretical-observational dichotomy). Here it is supposed that we can go inductively from $O_1(a)$ to $O_2(a)$ by first inducing $P(a)$ from $O_1(a)$ by (1.4) and then by deducing $O_2(a)$ from $P(a)$ on the basis of (1.5). If our above assumption which connects inductive inference with logical form is tenable, then in this argument we are implicitly relying on two principles of inductive inference. These principles are the Converse Entailment Condition and the Special Consequence Condition.

To state these conditions, let us write eIh for 'the statement h is *inducible* from the statement e'. This relation I of *inducibility* has been interpreted in various ways by the theorists of induction. In general, these interpretations are of the following two main types:

(1)   *eIh* if and only if *e* confirms (inductively supports) *h*.
(2)   *eIh* if and only if *h* may rationally be accepted on the basis of *e*.

Besides the proposed inductive rules of acceptance, the most usual technical explicates for the interpretations of *I* are the following two (cf. Carnap, 1962a):

(A)   *eIh* if and only if $P(h/e) > P(h)$ (the Positive Relevance Criterion)
(B)   *eIh* if and only if $P(h/e) \geqslant 1 - \varepsilon$, where $\varepsilon < 0.50$ (the High Probability Criterion).

Here *P* is a probability measure defined for the language in which *e* and *h* are stated. The nature and the interpretation of the relation *I* is discussed in later chapters in greater detail. It should be emphasized already here that one should not expect to find a unique explicate for *I*: different explicates of *I* may be appropriate within different inductive situations.

The above mentioned principles of inductive inference can now be stated as follows (to hold between any statements *h*, *e*, and *b*):

> *Converse Entailment*: If $e \vdash h$, then *hIe*.
> *Special Consequence*: If *eIh* and $h \vdash b$, then *eIb*.

Obviously, in inducing $P(a)$ from $O_1(a)$ by (1.4) we rely upon Converse Entailment (or some essentially similar principle), and when inferring $O_2(a)$ inductively from $O_1(a)$, (1.5), and from the fact that $O_1(a) IP(a)$, we use Special Consequence. However, it has been shown that these two general principles of inductive inference taken jointly lead to the absurd consequence that any statement is inducible from any other given statement (for a proof, see Hesse, 1970a). Furthermore, as has been demonstrated, the examples given by Hempel, Scheffler and others for showing the inductive indispensability of theoretical concepts rely either upon the joint use of the Converse Entailment Condition and the Special Consequence Condition, or then they rely upon some unacceptable transitivity principle (cf. Niiniluoto, 1973a).

Can the present dilemma, which may be dubbed *the transitivity dilemma*, be avoided? It seems that at least the following ways out could be tried.

(1) One may argue that inductive inference is not connected with the logical form of statements in the manner assumed above. For instance, one may claim that the inductive inference from the consequent of an implication-statement to its antecedent is not generally plausible even if it can be made in some special pragmatic contexts (such as in Hempel's example above).

(2) One may also argue that the Converse Entailment and Special Consequence conditions, and other relevant inductive principles relied on, have been incorrectly formulated. They should not hold for any arbitrary logical consequences between statements but only for those representing an *explanatory* relationship. Thus, in the formulation of these inductive principles, a relation E of explanation should be substituted for $\vdash$. This proposal has been explored to some extent in Tuomela (1973), and is also discussed in Chapter 11.4 below.

(3) One may argue that the inductive relations that occur in the above inductive principles are different (cf. Smokler, 1968). Thus, using Smokler's terminology, the Converse Entailment Condition holds true within 'abductive inference' while 'enumerative inference' satisfies the Special Consequence Condition.

(4) One may conceive of the inference from O_1 to O_2 *via P* not as an argument in two steps as above but as a *one-step* argument leading from the conjunction of theory T and $O_1(a)$ to $O_2(a)$, or from $O_1(a)$ to $O_2(a)$ 'in the light of T', that is, within the background framework of the theory in which P occurs (cf. Niiniluoto, 1973a).

Below, we shall show how the transitivity dilemma can be avoided by adopting the suggestion (4). By applying Hintikka's inductive logic, we give a general account of inductive systematization as established by one-step arguments, i.e., without relying on any problematic transitivity properties of induction. This account also serves to refute a number of other counterarguments to Hempel's thesis of the indispensability of theoretical concepts for inductive systematization (cf. Chapter 9).

1.2.   Before proceeding to questions of inductive systematization, for further reference we clarify what an *observational-theoretical dichotomy* means in this work.

Recent discussion concerning the classification of scientific concepts into observational and theoretical has made it clear that a *semantic* division can hardly be made (see, for example, Spector, 1966 and Achinstein, 1968). (This seems to be what the 'partial interpretation' theoreticians, for instance Carnap, the early Hempel, and Braithwaite, were looking for.) All scientific concepts seem to be semantically theory-laden, although they differ as to the extent of their theory-ladenness and as to what theories they are laden with. On the other hand, it can be argued that scientists do classify their concepts in various ways, and that some of these classifications are methodologically important. For instance, a dichotomy of concepts into (solely) *descriptive* and *explanatory* (but still partly descriptive) concepts is an important bifurcation of this type. To a great extent this dichotomy seems to go together with an 'intuitive' observational-theoretical classification (so far as that is explicable in general terms). Another useful bifurcation of concepts is that between (either historically or semantically) *old* versus *new* concepts. However, our emphasis will usually be on the observational-theoretical characterization.

In agreement with the above remarks, in this work we have adopted the following technical *methodological* characterization for an observational-theoretical dichotomy (originally proposed in Tuomela, 1973):

A nonlogical concept P occurring in a theory T belonging to a paradigm K is called *observational* with respect to theory T if and only if every representative scientist within K can (validly and reliably) 'measure' P in the typical applications of T without relying upon the truth of T.

A nonlogical concept P occurring in a theory T belonging to a paradigm K is called *theoretical* with respect to T if and only if (a) P is not observational (with respect to T) in the strong sense that it is true to say of every representative scientist within K that he cannot 'measure' P in *all* typical applications of T without relying upon the truth of T; and (b) P has been introduced into T in order to explain the behaviour (i.e., those aspects of it which are accounted for by T) of the objects T is about.

In this admittedly somewhat loose characterization the notion of paradigm is to be understood as a constellation of group commitments in the sense of Kuhn (1969). 'Measuring' P means coming to a conclusion whether or not atomic statements about P and some domain of objects hold true. There is no restriction as to the complexity of the measurement procedure.

The other notions that occur in the above characterization can for present purposes be understood in their presystematic sense (cf. Tuomela, 1973 for a discussion of them).

An interesting additional aspect of our theoretical-observational dichotomy has not been directly taken into account in the above explication, but is important for the developments in this work. This aspect is the possibility of using theoretical concepts, either directly or indirectly (by relying upon theory T), for reporting (i.e., 'evidentially'). (That is, evidential use of a theoretical concept is understood to be equivalent to its measurability in the above sense.) Since reliance upon the truth of the theory T is allowed, it is compatible with our characterization that theoretical concepts can never, sometimes, or always be used evidentially. In this book, we are concerned with the two extremes of this dimension. If a theoretical concept can *always* be used evidentially, we call it an *evidential* theoretical concept. If it can *never* be used evidentially, we call it a *non-evidential* theoretical concept.[1]

2. INDUCTIVE SYSTEMATIZATION ESTABLISHED BY THEORIES

2.1.   In this section, we discuss in general terms what inductive systematization is and what is meant by the logical indispensability of theoretical concepts for inductive systematization. A more detailed attempt at clarification of these notions can be found in Niiniluoto (1973a).

In the spirit of Hempel (1958), by 'systematization' we might mean any argument leading from a set of premises to a conclusion. In other words, this Hempelian notion of systematization involves some kind of inference from a set of statements (premises) to a statement (conclusion). If the premises of the systematizing argument are given, and a suitable conclusion is sought, we may speak of *positing* (in Scheffler's (1963) terminology). Obviously, scientific prediction is a subcase of positing. If again a possible conclusion is given, we may try to *substantiate* this conclusion by stating suitable premises from which the conclusion can in some sense be inferred. The varieties of scientific explanation (such as causal, reason-giving, etc.) are important subcases of substantiating arguments.

As the present investigation is focused upon the logical aspects of scientific systematization, we here classify cases of systematization by means of logical criteria – although in the hope of being able to capture something of philosophical interest. First, the statements in a systematization can be, by their logical nature, singular or (essentially) general. If the premises essentially contain lawlike statements or scientific laws (they have to be of essentially generalized form; cf. also Chapter 11.5) we call the systematization *nomological*; otherwise, it is anomological. In this book, only nomological systematization is discussed. Moreover, concentration will mainly be on systematizations of which the conclusions are general statements. Secondly, if the premises contain general probabilistic statements, we call the systematization *probabilistic*, otherwise it is non-probabilistic. The discussion below is mainly concerned with non-probabilistic systematizations in this sense of the word. (However, it is assumed that the statements in the systematizations can meaningfully be assigned inductive probabilities in the metalanguage.) The third and the most important logical aspect of scientific systematizations is the nature of the relationship that obtains between the premises and the conclusion of a systematization. If this relation is that of logical deduction, we speak of *deductive* systematization; otherwise, we speak of non-deductive systematization. The important subcase of non-deductive systematization concerned below is *inductive* systematization, viz., a systematization in which the premises of the argument stand in an inductive relationship to the conclusion.

Thus, the central topic of this book is *nomological, non-probabilistic inductive systematization of general statements.* As we do not require that the premises must be true, the systematizations to be discussed are only *potential* in Hempel's sense. In other words, we are concerned with what may be called *hypothetico-inductive inference* (cf. Chapter 11.2).

So far, we have been discussing scientific systematization as an argument of inference. This idea is obviously acceptable in the case of deductive systematization, but does not seem very plausible in the case of inductive systematization. Consider, for instance, a typical example of Hempelian inductive-probabilistic explanation:

$$P(G/F) = r$$
$$F(a)$$
$$\overline{\qquad\qquad} \; [r]$$
$$G(a)$$

Here the conclusion $G(a)$ is supposed to be inferable from the law $P(G/F)=r$ and the initial condition $F(a)$ with the inductive strength r (given in brackets). (Cf. Hempel, 1962.) Now, if the probability r in the law is very high we are in a fairly good position to inductively 'detach' $G(a)$ from the premises. But if r is small, say less than 0.50, we obviously cannot rationally expect $G(a)$ to occur on the basis of the premises. However, in some cases of this kind the initial probability of $G(a)$ may be very low in comparison with its probability relative to the premises. Then the premises are probabilistically relevant to $G(a)$, and may be regarded as having explanatory power with respect to $G(a)$. In information-theoretic terminology, it may then be said that the premises convey some information concerning $G(a)$, and thereby reduce the initial uncertainly associated with it.

Our viewpoint in this book is that inductive explanation, independently of whether the explanans is probabilistic or not, should be regarded as an information-providing argument but not as an argument for the inference or detachment, by some means, of a conclusion from some premises. Thus, to explain something is to give some information relevant to the explanandum. As we shall generally measure information in terms of probabilities in the usual fashion, it can alternatively be said that the explanation of something is the statement of some propositions probabilistically relevant to the explanandum (cf. Chapter 7). It is then only natural to adopt the Positive Relevance Criterion of the inducibility relation I (i.e., eIh iff $P(h/e)>P(h)$) for inductive explanations, and to proceed tentatively as if the strength of the inductive relationship between e and h were proportional to, if not identical with, $P(h/e)-P(h)$. In contrast, Hempel's idea of inductive explanation corresponds to the High Probability Criterion of the inducibility relation I (with a small ε) which may for instance be more plausible in the case of inductive predictions.

2.2. In our study, we are interested in systematizations which are established by scientific theories. By a theory we mean a deductively closed set of statements of some scientific language such that at least some of the statements are lawlike. However, below a theory will generally be identified with the conjunction of its axioms. Most of the technical discussion below is confined to theories axiomatized within monadic predicate logic (without identity). It will be assumed that the extralogical predicates of a theory have been dichotomized into a set λ of observational predicates, and a set μ of theoretical predicates in the sense explained in the preceding section. The full language of a theory will be denoted by $L(\lambda \cup \mu)$, or by L for short, and its sublanguage, containing only the members of λ as its extralogical predicates, by $L(\lambda)$.

Intuitively speaking, a theory establishes systematization if it appears as an essential premise of some systematization argument. Thus, a theory T establishes *deductive systematization* with respect to λ if and only if (i) $(T \& e) \vdash h$, and (ii) not $e \vdash h$ hold for some statements e and h of the observational language $L(\lambda)$. Similarly, we define:

A theory T in $L(\lambda \cup \mu)$ establishes *inductive systematization* with respect to λ and to the relation I if and only if for some statements k and h of $L(\lambda)$

(a) $(T \& k) I h$
(b)   not $k I h$
(c)   not $(T \& k) \vdash h$.

Here the statement k can be or can contain a statement that describes all the available evidence. For many purposes, however, it is better explicitly to relativize the above definition to an evidential situation described by e. Accordingly, we define:

A theory T in $L(\lambda \cup \mu)$ establishes inductive systematization with respect to λ and to the relation I, given the evidence e, if and only if for some statements k and h of $L(\lambda)$

(a′) $(T \& k \& e) I h$
(b′)   not $(k \& e) I h$
(c′)   not $(T \& k \& e) \vdash h$.

Our definition of inductive systematization is explicitly relativized to the relation I since there are conflicting explicates of I which may pertain to different kinds of inductive situations.

The condition (c) serves to separate the cases of deductive systematization from those of inductive systematization.

Our definition of inductive systematization established by a theory is ambiguous in several ways. For example, in conditions (a) and (b), I can be interpreted either as a rule of inference or as a binary predicate of the metalanguage or of the object language (cf. discussion in Niiniluoto, 1973a). Here, it is sufficient to point out the following important fact. The condition (b) – and similarly (b′) – allows of at least three different interpretations:

(a₁) h is inducible from the conjunction of T and k
(a₂) h is inducible from k relative to T
(a₃) T entails that h is inducible from k.

The difference between these interpretations can be illustrated by taking I to be the relation of positive relevance. Conditions (a₁)–(a₃) can then be expressed by

(A₁) $P(h/k \ \& \ T) > P(h)$
(A₂) $P(h/k \ \& \ T) > P(h/T)$
(A₃) $T \vdash [P(h/k > P(h)]$.

According to (A₁), '$k \ \& \ T$' is positively relevant to h. According to (A₂), k is positively relevant to h relative to T. Finally, according to (A₃), T entails that k is positively relevant to h. All these interpretations seem natural, and none of them implies another. In (a₁), the theory T is treated as a part of our total evidence, while (a₂) explicates inductive systematization 'seen in the light of the theory T'. (a₃) can be a suitable interpretation in the case that T is a probabilistic theory.

In terms of positive relevance, the condition (b) is expressed by

(B) $P(h/k) \leqslant P(h)$.

However, if (a) is interpreted as (A₃), then (b) needs to be interpreted as

(B′)   not $\vdash [P(h/k) > P(h)]$.

We are now in a position to define what is meant by the logical indispensability of theoretical concepts (members of μ) for inductive systematization. Corresponding to our unqualified notion of inductive systematization, we get:

The theoretical concepts μ of a theory T in $L(\lambda \cup \mu)$ are *logically indispensable* for inductive systematization with respect to λ and I if and only if whenever T establishes some inductive systematization with respect to λ and I there is no subtheory of T in $L(\lambda)$ which achieves at least the same inductive systematization.

An evidence-dependent notion of indispensability corresponding to our second notion of inductive systematization is obtained by relativizing both sides of this definition to e. A further absolute notion of logical indispensability is again obtained from this relativized notion, by stating that it suffices that the right hand side holds for some e.

Our definition of logical indispensability of theoretical concepts is motivated by the theoretician's dilemma argument, introduced by Hempel (1958). This instrumentalistic argument, which tries to show that theoretical concepts are unnecessary (dispensable) in science, depends on a premise to the effect that for any theory establishing deductive or in-ductive systematization with respect to λ we can effectively find a theory in $L(\lambda)$ that establishes the same systematization. In particular, it has been claimed that the well-known methods of Craig and Ramsey for the replacement of a theory in $L(\lambda \cup \mu)$ with another theory in $L(\lambda)$, having the same deductive consequences in $L(\lambda)$, suffice to show the dispensability of theoretical concepts in science (see, e.g., Cornman, 1972). If it can be shown – as we do in Chapter 9 – that the theoretical terms of some theory are logically indispensable for inductive systematization with respect to λ in the sense of our definition, this systematization is not preserved within Craigian or Ramseyan elimination, and the crucial premise of the theoreti-cian's dilemma is refuted.

Of course, there are other senses of indispensability of theoretical concepts which are stronger or weaker than our notion of logical in-dispensability. A number of philosophical reasons, for example those of ontological, semantical, and methodological nature, for the employment of theoretical concepts can be taken up in this connection (cf. Tuomela, 1973). In view of our emphasis on logical and methodological questions, the following remarks may suffice here.

It may happen that a theoretical term which is logically indispensable is *potentially dispensable* in the sense that it can, in future, be replaced by a new observational predicate. For example, 'gene', which was initially introduced into biology as a non-evidential theoretical term has now be-come an observational term of biology by virtue of the new strong in-struments we have at our disposal. However, whether a theoretical term is potentially dispensable or not is a factual question that depends upon the future development of scientific knowledge, and consequently cannot be decided on the basis of the kind of logical considerations employed in this book.

The logical indispensability of a theoretical term does not guarantee its methodological importance unless it occurs in a well-corroborated scientific theory. Moreover, even logically dispensable theoretical terms can have important methodological uses. It is possible to investigate the various gains obtainable by theoretical terms within deductive and inductive systematization with respect to λ – regardless of their logical indispensability or dispensability. Thus, theoretical terms can prove to be *methodologically indispensable* for systematization as well as for other general aims of scientific theorizing. As far as there are gains obtainable by the use of theoretical terms, we can say that they are *methodologically desirable* in science. Some of these gains, as measured by expected utilities, systematic power, and degrees of corroboration, are studied in Chapters 6–8 below.

3. A LOGICAL FRAMEWORK FOR THE DYNAMICS OF CONCEPTUAL CHANGE AND INDUCTION

3.1.   Conceptual change in science has long been a problem which has resisted exact philosophical, methodological, and logical analysis. In particular, philosophers have been puzzled by those cases of conceptual change, or at least conceptual growth, which involve the introduction of theoretical concepts. It needs to be added that the important point about conceptual change and growth is of course that they provide additional potentialities for the expression and procession of information not expressible within the original conceptual framework. When problems of conceptual enrichment are placed in the context of scientific inference, especially *inductive* inference, new difficulties undoubtedly arise. In this work, we attempt to consider jointly these two major philosophical problems, conceptual change and induction, and try to investigate their interplay in some simple contexts.

Let us first characterize in general terms what is meant by conceptual growth and inductive change, and how they will be jointly accounted for and treated in this work. In this section we concentrate on issues which are central from a philosophical aspect. The technical details will be given in the appropriate contexts of later chapters.

Consider first a scientific language $L(\lambda)$ with λ as its set of nonlogical predicates. We normally understand that the members of λ are observational, but sometimes it is useful to think of them just as 'old' predicates. $L(\lambda)$ is the language in which the scientist originally states his evidence and his generalizations. Next consider an enriched language $L(\lambda \cup \mu)$ obtained by adding a set μ of theoretical (or 'new') predicates to $L(\lambda)$. The adoption of $L(\lambda \cup \mu)$ instead of $L(\lambda)$ means potentially more for a scientist than does simply linguistic enrichment. For he may now be able to describe the world in a deeper and richer way and to process information not expressible without the theoretical concepts of μ.

In terms of induction, we can simply put this in the following way. Let P_0 be a conditional measure of inductive probability defined for the statements of $L(\lambda)$. The measure P_0 then only takes into account observational evidence and can be used to measure the inductive support, etc., of only observational statements such as generalizations. However, if we proceed from $L(\lambda)$ to $L(\lambda \cup \mu)$ and are able to define an adequate probability measure P_1 for the statements of $L(\lambda \cup \mu)$, we can avoid these restrictions. For such a measure will be defined for all statements (such as generalizations and theories), no matter whether they are statable *only* in $L(\lambda \cup \mu)$, or whether they can be stated in $L(\lambda)$. Similarly, the evidence may contain not only observational evidence (e), but also several kinds of background information (T). (As we shall see, T may be or contain a proper scientific theory, semantic postulate, auxiliary hypothesis, etc.) If a generalization g and a piece of evidence e have originally been stated in $L(\lambda)$, in a sense we reinterpret them within $L(\lambda \cup \mu)$; see below and Chapters 3 and 4 for this reinterpretation.

In all, we are here dealing with the following three types of factors:

(1)   *Conceptual growth* as characterized by the shift from $L(\lambda)$ to $L(\lambda \cup \mu)$;

(2)   *Inductive change* as characterized by the adoption of the measure P_1 (instead of P_0).

(3)   The effect of *background information* taken into account by a (possibly complex) statement T, which may be expressible only with the help of theoretical concepts.

A number of methodological situations can be characterized in these terms. For instance, we may compare the probabilities $P_0(g/e)$ in $L(\lambda)$ with the new probabilities $P_1(g/e \& T)$ (in $L(\lambda \cup \mu)$) which take into account the background information T. It goes without saying that a variety of measures of inductive or factual support, corroboration, explanatory power, etc., which are definable on the basis of probabilities, can be analogously compared for the effect of conceptual change and background information.

Once we have the probabilities $P_0(g/e)$ and $P_1(g/e \& T)$, it may be asked whether T establishes inductive systematization with respect to $L(\lambda)$ (in the sense of our definition in Section 2). Similarly the question about the logical indispensability (for inductive systematization), and that about the methodological desirability of the theoretical concepts of T, can be investigated. Later the role of theoretical concepts as to their indispensability and desirability for inductive systematization and explanation is discussed rather extensively (cf. in particular Chapters 6–9).

Let us briefly compare classical *hypothetico-deductive* inference with our approach (cf. Chapter 11 for fuller treatment). Assume that T is a scientific theory in $L(\lambda \cup \mu)$ which entails an observational generalization g. Consider now some observational evidence e. If e contradicts g, T is logically refuted (in so far as e expresses reliable evidence). If e inductively supports or corroborates g, then e has also to support T. If again T does not logically entail g but only gives 'extraobservational' or theoretical background support to g we are dealing with a situation which will be termed *hypothetico-inductive* inference. This kind of situation may, but need not, reflect our ignorance: greater knowledge might enable us to find a stronger T^* such that $T^* \vdash g$. Analogously with the hypothetico-deductive case, evidence that gives support to g, or evidence that disconfirms g, must have some effect upon the degree of corroboration or inductive support of T. The form of this effect may depend in part upon the nature of the theoretical concepts that occur in T.

Another aspect of the logic of hypothetico-inductive inference is the question of how (hypothetical) theory T contributes to the establishment of inductive systematization with respect to $L(\lambda)$. In other words, instead of studying the inductive support, etc., of theory T itself, we can investigate the effects of T upon the inductive status of generalizations g in $L(\lambda)$. It is with this latter aspect of hypothetico-inductive inference that we are mainly concerned in this book, even if some comments on the former aspect are be made in Chapters 8 and 11.

3.2.   Now we state and discuss more precisely seven conditions of adequacy for a probability measure P_1 which is supposed to take into account conceptual and inductive change as well as the effect of background information.

(CA1) P_1 is a (conditional) probability measure defined for the statements of a language $L(\lambda \cup \mu)$ in which λ is a set of observational predicate constants, and μ a set of theoretical predicate constants.

In other words, P_1 should be such that we can have probabilities $P_1(g/e \; \& \; T)$ where g, e, and T are in $L(\lambda \cup \mu)$; in particular, T may not be expressible within $L(\lambda)$. CA1 thus involves that P_1 is defined so as to account for both the observational evidence e and the (theoretical) background information T. (Below we discuss in greater detail the nature of this background information.) CA1 can, furthermore, be understood in the strong sense that P_1 will depend upon the logical structure of the theory T. It would then be possible that the value of P_1 depends upon the degree to which the theoretical concepts of T are definable or involved with the observational predicates in T (cf. Chapter 5).

Once we have such a measure P_1, we can compute the factual or inductive support, the degree of corrroboration, etc., of observational generalizations and the explanatory power of theories (containing theoretical predicates) with respect to them, provided that these properties are analyzable in terms of probabilities (cf. Chapters 6, 7, and 8).

(CA2)   It should be possible to have at least asymptotically (with increasing evidence e) $P_1(g/e \; \& \; T) = 1$ where g is a generalization, e expresses observational evidence, and T is a background theory.

If (and perhaps only if) CA2 is satisfied, it seems practicable to define measures for the degree of corroboration or inductive support which serve to estimate the degree of truth (or closeness to truth) of theories (with theoretical concepts). (Cf. Chapter 8.)

Next we have a general requirement concerning the dynamics of inductive change within $L(\lambda \cup \mu)$:

(CA3) P_1 should 'learn from experience' both in an enumerative and eliminative sense.

By CA3 we mean roughly that the measure P_1 should be a function of the amount of evidence (cf. CA2). At the same time when mere enumeration of evidence changes the values of P_1, P_1 should inductively discriminate (and select between) competing generalizations on the basis of the observational evidence *cum* background information. In particular, we shall understand the requirement of elimination in the sense of the following principle, to be termed the principle of *inductive epistemic orderliness*: The probability mass should concentrate on fewer and fewer competing generalizations with growing evidence, provided that this evidence is sufficiently varied. In other words, it is required that, within our richer conceptual framework, induction should bring about some orderliness among generalizations so that we obtain some clues as to which generalizations might be true, and which false.

Next, we have a requirement concerning the inductive effectiveness of the background information T.

(CA4)   It should be possible that the background information is inductively effective (both preasymptotically and asymptotically).

CA5 will be so understood so that it is possible that $P_1(g/e \ \& \ T) \neq P_0(g/e)$ for finite evidence. Furthermore, it should be possible that the inductive support of g given by e and T approaches its maximum while the support of g given by e alone approaches its minimum when the amount of evidence grows without limit.

Following this, we have a condition concerning the dynamics of conceptual change from $L(\lambda)$ to $L(\lambda \cup \mu)$:

(CA5)   The measure P_1 should be such that it is possible to account for both linguistic variance and linguistic invariance.

According to this requirement, P_1 should be able to reflect inductive changes which arise from 'non-conservative' shifts of conceptual systems. The content of CA5 will be explained in detail in Chapter 10.

(CA6)   Probability measure P_1 should not be determined on purely *a priori* or logical grounds.

This requirement should be understood in the sense that metalinguistic probability statements of the form '$P_1(g/e \ \& \ T)=r$' are partly factual. Below, we employ Hintikka's two-parametric system of inductive logic, in which the values of the two parameters cannot be fixed on purely *a priori* or logical grounds.

As our last requirement, we have a somewhat more special condition related to the transitivity dilemma (cf. Section 1), which arose from the lack of reasonable transitivity properties of the inducibility relation.

(CA7) P_1 should be such that the transitivity dilemma is avoidable.

As will be seen in Chapters 9 and 11, the transitivity dilemma does not arise within our framework, essentially because no need exists longer for the use of mutually incompatible principles of inductive inference.

3.3.   What we plan to accomplish in all theoretical detail is the definition of a probability measure P_1 – satisfying CA1–CA7 – for *monadic* languages $L(\lambda \cup \mu)$. For this purpose, Hintikka's parametric systems of inductive logic will be applied. Thus, even if our concepts and methods are defined in general terms, the technical treatment has been carried out only for monadic situations. Even if this is a clear restriction (from both logical and methodological viewpoints) the reader is reminded that most of the technical discussion concerned with problems of induction – such as the transitivity dilemma – has been conducted by means of monadic examples. If one wishes to solve the puzzles and paradoxes resulting from that discussion, those same monadic examples obviously need to be treated.

Within the monadic framework employed in most of this work, we have as initially possible hypotheses all the essentially *general* statements that can be formulated by means of the monadic language in question (be it $L(\lambda)$ or $L(\lambda \cup \mu)$). Similarly, it is assumed that our background information (T) is given in the form of general statements. Mostly, it will be assumed that T can be expressed *only* by using theoretical concepts, i.e., by means of $L(\lambda \cup \mu)$.

Many of the methodological situations to be discussed in this work involve comparison of the probabilities $P_1(g/e \ \& \ T)$ (in $L(\lambda \cup \mu)$) and

$P_0(g/e)$ (in $L(\lambda)$). (That this will be the case is obvious, in view of our interest in positive inductive relevance.) Usually we assume that g is an observational generalization originally expressed in $L(\lambda)$. In shifting from $L(\lambda)$ to $L(\lambda \cup \mu)$ g is 'reinterpreted' or 'redescribed' within $L(\lambda \cup \mu)$ (see Chapter 3). On the other hand, T is not expressible within the observational language $L(\lambda)$ at all, even if it may have (purely) observational consequences. The background information T may have quite a few methodological functions, some of which are indicated below.

First of all, T may be or contain a proper scientific theory, which is more or less hypothetically accepted for explanation, prediction, and, of course, further testing. Secondly, T may contain auxiliary scientific hypotheses or theories, statements about experimental conditions, and the like. Thirdly, T may express semantic postulates and conventions regulating the use of the predicates of $L(\lambda \cup \mu)$. It may also express some *a priori* metaphysical assumptions (e.g. a principle of causality) in so far as these assumptions are given in a form sufficiently specific for them to be statable within $L(\lambda \cup \mu)$.

The methodological role (or, rather, roles) of T will be discussed in later chapters. At the moment, it is sufficient to indicate a few of them. Within the contexts of hypothetico-inductive (and hypothetico-deductive) inference, T can play the roles of an explanans and a predictor (cf. Chapter 7). It can further serve to provide theoretical support (support coming from 'upward', from a 'nomological network' of theories) to (observational) generalizations g (see Chapter 8). In many of these cases, the relevant inductive probabilities of any other statements concerned are relativized to the truth to T; hence these other statements are 'seen in the light of T'. (This is in the spirit of scientific realism.) However, T can often also – somewhat instrumentalistically – play the role of 'theoretical' evidence treated on a par with the purely observational evidence e.

In terms of scientific pragmatics, a scientist trying to maximize the epistemic utility (information content) of his rivalling hypotheses will often find the role of a background theory crucially important (see Chapter 6).

Some technical functions of T follow. First, T can be employed to define the set of rivalling hypotheses in any given methodological situation. That is, T may be used to select such a set out of the set of all logically possible hypotheses compatible with the evidence. In our system, this may markedly affect the probabilities of the individual hypotheses (cf. Chapters 3 and 4). Secondly, T may be employed to create asymmetries between complex predicates (cf. Chapter 11.5). This feature is of help in the analysis of notions such as natural kinds or projectible predicates, and so on.

So far, very little has been said about the evidence e for our theories and hypotheses. It can be given in various ways. First, the problem arises of whether the evidence can be described only by means of $L(\lambda)$, or whether statements of $L(\lambda \cup \mu)$ can always, or at least sometimes, be employed for this purpose. (This naturally hinges upon one's view of the nature of theoretical concepts; cf. Section 1) Secondly, there is the question of whether the evidence can be given by *complete* descriptions of individuals. Within a monadic language, each complex predicate (Ct-predicate in our, and Q-predicate in Carnap's terminology) is defined as a conjunction in which all the primitive predicates, or negations of them, occur. Incomplete evidence statements are then statements which attribute only disjunctions of Ct-predicates to observed individuals. Thirdly, the evidence statements can be 'deterministic' (definite) or probabilistic. In the deterministic case, evidence statements are of the form '$F(a)$' (or, maybe rather, "'$F(a)$' is (provisionally) accepted as true") and the probabilistic case gives evidence statements of the form '$P(F(x)) = r$'. Something is said below about the first two of these three aspects.

In accordance with our definition of inductive systematization and the logical indispensability of theoretical concepts, we are below mainly concerned with the changes in the inductive probabilities of *generalizations* arising from new theoretical concepts and new information encoded in the language $L(\lambda \cup \mu)$. Our view is that one of the main goals of scientific activity is that of seeking *general* knowledge, rather than the mere collection of singular data, and perhaps the systematization of such data. In other words, (pure) science looks for general patterns and regularities in the world; only applied science and technology take a primary interest in singular events and facts. Somewhat surprisingly, this obvious point has not been appreciated by most contemporary analytical philosophers; in general, they have dealt only with questions such as those of systematizing, inferring and explaining singular data.

In the spirit of our philosophical orientation, no systematic discussion will be made of the inductive systematization, etc., of singular statements in this book. (Furthermore, the probabilities of singular statements in Hintikka's inductive logic markedly depend upon the probabilities of generalizations; treating the probabilities of singular statements can be minimized also for this reason.)

From what has been said above, it can be inferred that our general sympathies are for *scientific realism*, and against *scientific instrumentalism*. Generally speaking our (critical) scientific realism claims that science is about a reality that exists independently of observers. In principle this reality is knowable, though in a symbolic and distorted way. Knowledge about this reality is primarily obtained by means of scientific theorizing and experimentation, and this knowledge always remains corrigible. The important point about this kind of factual knowledge is that it need not be totally empirical (sensible *stricto* or *lato sensu*), or even have a clear empirical correlate. It is primarily, but not merely, for the attainment of such nonempirical factual knowledge that theoretical concepts are needed. (Cf. Tuomela (1973) for a more detailed discussion of critical scientific realism.)

It follows that a realist may conceive of scientific theories as (genuinely) true or false, whereas an instrumentalist regards them at best as methodologically convenient instruments or tools for systematizing observational statements. The main philosophical part of this book consists of what is hoped to be a fair attempt to show that – as opposed to some recent claims – instrumentalism is an unviable doctrine both philosophically and methodologically.

NOTE

[1] Sellars (1965) has argued that a scientific concept does not have full-fledged *existential* status (in the sense that it can refer to existing objects) unless it has at least some *direct* evidential uses. As a (Sellarsian) scientific realist – in distinction to a scientific instrumentalist – normally wants his theoretical concepts to have such a full existential status, within our framework theoretical concepts having at least some direct evidential uses come to reflect scientific realism, while especially non-evidential ones go together with instrumentalism. It should be noted, however, that even if a theoretical concept has no (direct or indirect) evidential uses, it may still be a referring concept as to its semantical status.

CHAPTER 2

HINTIKKA'S TWO-DIMENSIONAL CONTINUUM
OF INDUCTIVE METHODS

In this study, Hintikka's inductive logic will be used as a general framework for discussion of the technical problems of inductive systematization. For later reference, this chapter summarizes some basic results about Hintikka's two-dimensional continuum of inductive methods.[1]

1. SUMMARY OF HINTIKKA'S TWO-DIMENSIONAL CONTINUUM

Let L_N^k be an applied monadic first-order language with k primitive predicates $P_i (i=1, 2, ..., k)$ and N individual constants $a_i (i=1, 2, ..., N)$. Conjunctions of the form

$$(2.1) \qquad Ct_j(x) = (\pm) P_1(x) \& (\pm) P_2(x) \& ... \& (\pm) P_k(x),$$

where the symbol $(\pm)$ may be replaced by a negation sign or nothing, will be called the *attributive constituents* or the *Ct-predicates* of L_N^k. The number of different Ct-predicates of L_N^k is $K = 2^k$. The Ct-predicates $Ct_j(x) (j=1, 2, ..., K)$ describe all different kinds of individuals which can be specified by the language L_N^k. Conjunctions of the form

$$(2.2) \qquad (\pm) (Ex) Ct_1(x) \& (\pm) (Ex) Ct_2(x) \& ... \& (\pm) (Ex) Ct_K(x),$$

where it is specified for each Ct-predicate whether it is instantiated in the universe U or not, will be called the *constituents* of L_N^k. The constituent (2.2) can be written alternatively in a form which enumerates all the existing kinds of individuals and adds that other kinds of individuals do not exist:

$$(2.3) \qquad (Ex) Ct_{i_1}(x) \& (Ex) Ct_{i_2}(x) \& ... \& (Ex) Ct_{i_w} \&$$
$$\& (x) (Ct_{i_1}(x) \lor Ct_{i_2}(x) \lor ... \lor Ct_{i_w}(x)).$$

The constituent (2.3) is referred to as C_w. The number of different constituents of L_N^k is 2^K. They describe all the different kinds of 'possible worlds' that can be specified by means of our k primitive predicates, sentential connectives and quantifiers. All general sentences of L_N^k are transformable into a disjunction of constituents. This disjunction is called the *distributive normal form* of the sentence in question. General sentences which contain only one constituent in their distributive normal form are called *strong generalizations*. Other generalizations are called *weak*.

Suppose now that we have observed n different individuals $a_i (i = 1, 2, ..., n)$ of the universe U, and that we have observed them completely in the sense that the Ct-predicate of L_N^k which a_i exemplifies is known for all a_i. Let the exemplified Ct-predicates be $Ct_{i_1}, Ct_{i_2}, ..., Ct_{i_c}$, and let the numbers of observed individuals in these classes be $n_1, n_2, ..., n_c$, respectively. Then c is the number of different kinds of individuals we have observed, and the sum of $n_j (j = 1, 2, ..., c)$ is n. Let e be a singular sentence which describes this sample.

A constituent C_w is compatible with our evidence e if $c \leqslant w \leqslant K$. Hence, by Bayes's formula,

$$(2.4) \qquad P(C_w/e) = \frac{P(C_w) P(e/C_w)}{\sum\limits_{i=0}^{K-c} \binom{K-c}{i} P(C_{c+i}) P(e/C_{c+i})}.$$

To determine the value of the *a posteriori* probability $P(C_w/e)$ of C_w, we have to specify the *a priori* probabilities $P(C_w)$ and the likelihoods $P(e/C_w)$. In the following, we shall describe how these values are determined in Hintikka's two-dimensional system of inductive logic. It will be assumed that the universe U is infinite, i.e., that $N \to \infty$.

The representative function of Carnap's λ-system of inductive logic is

$$\frac{n_j + \lambda/K}{n + \lambda}.$$

(Cf. Carnap, 1952a.) In Hintikka's system, it is replaced by the representative function (conditional on the constituent C_w)

$$(2.5) \qquad \frac{n_j + \lambda(w)/w}{n + \lambda(w)}$$

which gives the probability of the hypothesis $Ct_{i_j}(a_{n+1})$ where a_{n+1} is some unobserved individual, n_j is the number of individuals in our sample e exemplifying the Ct-predicate Ct_{i_j}, w the number of Ct-predicates instantiated in our universe U according to constituent C_w, and λ a parameter ($0 \leqslant \lambda \leqslant \infty$) possibly depending on w. By (2.5) we then have

$$(2.6) \qquad P(e/C_w) = \frac{\Gamma(\lambda(w))}{\Gamma(n + \lambda(w))} \prod_{j=1}^{c} \frac{\Gamma(n_j + \lambda(w)/w)}{\Gamma(\lambda(w)/w)}$$

where Γ is the Gamma-function. (Note that $\Gamma(n+1) = n!$.)

In Carnap's λ-system, the probability that a set of α individuals is compatible with C_w is

$$\frac{\Gamma(\alpha + w\lambda/K)\, \Gamma(\lambda)}{\Gamma(\alpha + \lambda)\, \Gamma(w\lambda/K)}.$$

Hintikka (1966) proposes that the *a priori* probability $P(C_w)$ should be chosen to be proportional to this probability. Thus, in Hintikka's system

$$(2.7) \qquad P(C_w) = \frac{\dfrac{\Gamma(\alpha + w\lambda(K)/K)}{\Gamma(w\lambda(K)/K)}}{\displaystyle\sum_{i=0}^{K} \binom{K}{i} \dfrac{\Gamma(\alpha + i\lambda(K)/K)}{\Gamma(i\lambda(K)/K)}}.$$

Then $P(e)$, i.e., the denominator of formula (2.4), by (2.6) and (2.7) becomes

$$(2.8) \qquad P(e) = \frac{\displaystyle\sum_{i=0}^{K-c} \left\{ \binom{K-c}{i} \dfrac{\Gamma(\alpha + (c+i)\,\lambda(K)/K)}{\Gamma((c+i)\,\lambda(K)/K)} \times \right.}{\displaystyle\sum_{i=0}^{K} \binom{K}{i} \dfrac{\Gamma(\alpha + i\lambda(K)/K)}{\Gamma(i\lambda(K)/K)}}$$

$$\left. \times \dfrac{\Gamma(\lambda(c+i))}{\Gamma(n + \lambda(c+i))} \prod_{j=1}^{c} \dfrac{\Gamma(n_j + \lambda(c+i)/(c+i))}{\Gamma(\lambda(c+i)/(c+i))} \right\}$$

The probability $P(C_w/e)$ can then be computed by formulae (2.4), (2.6), (2.7) and (2.8). These formulae define Hintikka's two-dimensional continuum of inductive methods (with the non-negative real-valued parameters λ and α).

The *a priori* probability of constituents was so chosen that, the greater α is, the greater is $P(C_K)$, i.e., the *a priori* probability of the constituent C_K according to which there are all kinds of individuals in our universe. When $\alpha \rightarrow \infty$, $P(C_K)$ approaches one, and the probabilities of all the other constituents of our language approach zero. In this case, we have an 'atomistic' universe corresponding to Carnap's λ-system. If $\alpha = 0$, then all constituents have equal *a priori* probabilities. Consequently, the parameter α can be thought of as "an index of the strength of *a priori* considerations in inductive generalization" (cf. Hintikka, 1966, p. 117). As such, it plays the role of "an index of caution" within inductive generalization (cf. Hintikka, 1970, p. 21). The greater α is, the less regularly the individuals of our universe have a tendency to concentrate into few Ct-predicates, and the less probable are universal laws in our universe. Thus, speaking in objectivistic terms, α can be thought of as expressing the amount of disorder or irregularity that there probably exists in the universe as far as general laws are concerned, or, in subjectivistic terms, as representing the expectations of the investigator in regard to the amount of this disorder. In this respect, α is comparable to the parameter λ, which can be thought of as a measure of the disorder of the universe, or our belief in this disorder, as is expressed by singular sentences concerning the distribution of individuals between those Ct-predicates that are instatiated in our universe. The difference between these parameters is that α is relevant to inductive generalization, and λ to singular inference (cf. Hintikka, 1966, p. 118; 1970, p. 21). As the parameters α and λ are extralogical, if follows that the adequacy condition CA6 of Chapter 1.3 is satisfied.

The basic result, which holds for all probability measures in the two-parametric system with $\alpha \neq \infty$ and $\lambda > 0$, is that when the size of our sample (i.e., n) grows without limit, the value $P(C_c/e)$ approaches one, i.e.,

$$\lim_{n \rightarrow \infty} P(C_c/e) = 1$$

$$\lim_{n \rightarrow \infty} P(C_w/e) = 0, \text{ if } w > c.$$

Here, C_c is the constituent according to which there exists in our universe U only such kinds of individuals that are exemplified in our sample. Thus,

Hintikka's system satisfies the adequacy conditions CA2 and CA3 of Chapter 1.3.

In our study, we are mostly interested in the probabilities of general sentences. Accordingly, for simplicity, we shall usually consider only a special case of Hintikka's two-dimensional continuum which is a kind of generalization of Carnap's confirmation function c^*. Following Hintikka (1968a) and Hilpinen (1968), we choose the parameter λ in such a way that $\lambda(w) = w$. This choice is, in fact, only a way of normalizing singular inference, and does not affect the results in regard to inductive generalization that are of primary interest to us. In this case, formulae (2.5)–(2.8) reduce to

$$(2.5')\qquad \frac{n_j + 1}{n + w}$$

$$(2.6')\qquad P(e/C_w) = \frac{(w - 1)!}{(n + w - 1)!} \prod_{j=1}^{c} (n_j!)$$

$$(2.7')\qquad P(C_w) = \frac{\dfrac{(\alpha + w - 1)!}{(w - 1)!}}{\displaystyle\sum_{i=0}^{K} \binom{K}{i} \dfrac{(\alpha + i - 1)!}{(i - 1)!}}$$

$$(2.8')\qquad P(e) = \frac{\displaystyle\sum_{i=0}^{K-c} \binom{K - c}{i} \dfrac{(\alpha + c + i - 1)!}{(n + c + i - 1)!} \prod_{j=1}^{c} (n_j!)}{\displaystyle\sum_{i=0}^{K} \binom{K}{i} \dfrac{(\alpha + i - 1)!}{(i - 1)!}}.$$

By formulae (2.4), (2.6'), (2.7'), and (2.8'), we now get

$$(2.9)\qquad P(C_w/e) = \frac{1}{\displaystyle\sum_{i=0}^{K-c} \binom{K - c}{i} \dfrac{(\alpha + c + i - 1)! \, (n + w - 1)!}{(n + c + i - 1)! \, (\alpha + w - 1)!}}.$$

If we put $\alpha = 0$ in these formulas, results are obtained for Hintikka's 'combined system' (cf. Hintikka, 1965b). In this case, we completely rely upon *a posteriori* considerations as far as inductive generalization is concerned. If we put $\alpha = 0$, and let $\lambda \to \infty$, we obtain Hintikka's 'Jerusalem system' (cf. Hintikka, 1965a). In this case, the formulae (2.5), (2.7), and (2.9) reduce to

(2.5″) $\dfrac{1}{w}$

(2.7″) $P(C_w) = \dfrac{1}{2^K}$

(2.9″) $P(C_w/e) = \dfrac{1}{\displaystyle\sum_{i=0}^{K-c} \binom{K-c}{i}\left(\dfrac{w}{c+i}\right)^n}$.

The probabilities of all generalizations can be computed by the formulae given above, viz., if g is a generalization such that its distributive normal form contains the constituents $C_{i_1}, C_{i_2}, ..., C_{i_g}$, then

$$P(g/e) = \sum_{j=i_1}^{i_g} P(C_j/e).$$

2. THE TREATMENT OF INCOMPLETE EVIDENCE

In our summary of Hintikka's system, it has been assumed that the evidence e is *complete* with respect to the language L_N^k, i.e., we assumed that for each individual in the evidence e, it is known which Ct-predicate of L_N^k it satisfies. In the treatment of problems of inductive systematization, we have simultaneously to consider two languages, with one of them a sublanguage of the other. In these cases, statements of the poorer language can be treated as disjunctions of statements of the richer language. In particular, evidence statements that are complete with respect to the poorer language may, in some cases, be incomplete with respect to the richer language. To handle these cases of incomplete evidence, we need a generalization of formula (2.5′).

Let $M_1, M_2, ..., M_h$ be predicates of L_N^k of width 2, i.e., every M_j is a disjunction of two Ct-predicates $Ct_{i_j}^1$ and $Ct_{i_j}^2$ ($j = 1, 2, ..., h$). Let e be a singular sentence describing a sample of n individuals such that the number of individuals in the classes $M_1, M_2, ..., M_h$ is $n_1, n_2, ..., n_h$, respectively. (This evidence e is incomplete in the sense that it is not assumed that we know which Ct-predicates the sampled individuals exemplify, but only that they satisfy a disjunction of two Ct-predicates.) Then for an unobserved individual a_{n+1},

$$(2.10) \quad P(M_i(a_{n+1})/e) = \frac{n_i + 2}{n + w},$$

in Hintikka's system with $\lambda(w) = w$.

For the proof of (2.10), note first that the probability of a conjunction according to which $n_1 - j_1$ individuals have the Ct-predicate $Ct_{i_1}^1$, j_1 have $Ct_{i_1}^2$, $n_2 - j_2$ have $Ct_{i_2}^1$, j_2 have $Ct_{i_2}^2$, ..., $n_h - j_h$ have $Ct_{i_h}^1$, and j_h have $Ct_{i_h}^2$ is by (2.5')

$$\frac{1 \cdot 2 \cdot \cdots \cdot (n_1 - j_1) \cdot 1 \cdot 2 \cdot \cdots \cdot j_1}{w(1 + w) \ldots (n_1 - j_1 - 1 + w)(n_1 - j_1 + w) \ldots (n_1 - 1 + w)} \times$$

$$\times \frac{1 \cdot 2 \cdot \cdots \cdot (n_2 - j_2) \cdot 1 \cdot 2 \cdot \cdots \cdot j_2}{(n_1 + w) \ldots (n_1 + n_2 - 1 + w)} \cdots \frac{j_h}{(n - 1 + w)}$$

$$= \frac{(w - 1)!}{(n - 1 + w)!} (n_1 - j_1)! j_1! (n_2 - j_2)! j_2! \ldots (n_h - j_h)! j_h!$$

for all $j_1 = 0, \ldots, n_1$; $j_2 = 0, \ldots, n_2$; ...; $j_h = 0, \ldots, n_h$. The sentence

$$A = M_1(a_1) \& \ldots \& M_1(a_{n_1}) \& M_2(a_{n_1+1}) \& \ldots \&$$
$$\& \ldots \& M_2(a_{n_1+n_2}) \& \ldots \& M_h(a_n)$$

is equivalent to a disjunction of 2^n conjunctions $\binom{n_1}{j_1}\binom{n_2}{j_2}\ldots\binom{n_h}{j_h}$ of which have the probability computed above. Hence,

$$P(A) = \sum_{j_1=0}^{n_1} \sum_{j_2=0}^{n_2} \cdots \sum_{j_h=0}^{n_h} \binom{n_1}{j_1}\binom{n_2}{j_2}\ldots\binom{n_h}{j_h} \times$$

$$\times \frac{(w - 1)!}{(n - 1 + w)!} (n_1 - j_1)! j_1! \ldots (n_h - j_h)! j_h!$$

$$= \sum_{j_1=0}^{n_1} \sum_{j_2=0}^{n_2} \cdots \sum_{j_h=0}^{n_h} \frac{(w - 1)!}{(n - 1 + w)!} n_1! n_2! \ldots n_h!$$

$$= \frac{(w - 1)!}{(n - 1 + w)!} (n_1 + 1)!(n_2 + 1)! \ldots (n_h + 1)!$$

Similarly, we have

$$P(A \ \& \ M_i(a_{n+1})) = \frac{(w-1)!}{(n+w)!} (n_i + 2) \prod_{j=0}^{h} (n_j + 1)!$$

Hence,

$$P(M_i(a_{n+1})/e) = \frac{P(A \ \& \ M_i(a_{n+1}))}{P(A)} = \frac{n_i + 2}{n + w}.$$

As a particular case of formula (2.10), we have

$$(2.11) \quad P(M(a_{n+1})/M(a_1) \ \& \ \cdots \ \& \ M(a_n)) = \frac{n+2}{n+w},$$

where M is any predicate of width 2.

NOTE

[1] Hintikka's system of inductive logic is developed in Hintikka (1965a, 1965b, and 1966). Its relations to enumerative and eliminative aspects of induction is discussed in Hintikka (1968a); to the rules of acceptance in Hintikka and Hilpinen (1966), Hilpinen (1968, 1972); to the theory of semantic information in Hintikka and Pietarinen (1966), Hintikka (1968b, 1970); and to the paradoxes of confirmation in Hintikka (1969a). Extensions of Hintikka's system are given in Hilpinen (1966), Tuomela (1966), and Hilpinen (1971). Further discussion about Hintikka's system is provided by Hintikka (1969b, 1971a), and Hilpinen and Hintikka (1971). For studies employing Hintikka's inductive logic, see Pietarinen (1972) and Uchii (1972).

CHAPTER 3

INDUCTIVE PROBABILITIES
OF WEAK GENERALIZATIONS

1. Probabilities in the Observational Language

In this chapter, we are interested in the question of how the probabilities of weak generalizations of the observational language change when a new predicate is introduced into our language, and, at the same time, our knowledge of the connections of this new predicate to the old vocabulary is applied. The same question for strong generalizations is taken up in the next section. General formulae are calculated for the probabilities of generalizations in Hintikka's system with $\lambda(w) = w$, and some observations and examples of their behaviour are given. For simplicity, we generally consider only the case of one new monadic predicate.

Let $L(\lambda)$ be the observational (or, old) language with the primitive predicates $\lambda = \{O_1, O_2, ..., O_k\}$. Let the Ct-predicates of $L(\lambda)$ be $Ct_1, Ct_2, ..., Ct_K$, where $K = 2^k$. Let e be a singular sentence completely describing a sample of n individuals in terms of the Ct-predicates of $L(\lambda)$, and let $Ct_{i_1}, Ct_{i_2}, ..., Ct_{i_e}$ be the Ct-predicates exemplified in this sample. Let g be a general sentence of $L(\lambda)$, i.e., an observational generalization, which is compatible with evidence e, and states that at least certain b Ct-predicates of $L(\lambda)$ are empty. We shall first compute the probability $P(g/e)$.

Generalization g is equivalent to a disjunction of constituents of $L(\lambda)$. Some of these constituents may be incompatible with evidence e. Suppose that g is equivalent to the disjunction of the constituents $C_{j_1}, C_{j_2}, ..., C_{j_g}$, when e is assumed, that is to say,

$$e \vdash g \equiv C_{j_1} \lor C_{j_2} \lor ... \lor C_{j_g}.$$

Then

(3.1) $\quad P(g/e) = \sum_{i=j_1}^{j_g} P(C_i/e) = \frac{1}{P(e)} \sum_{i=j_1}^{j_g} P(C_i) P(e/C_i).$

Let, for all $i = 1, 2, ..., g$,

$w_i =$ the number of Ct-predicates of $L(\lambda)$ instantiated according to the constituent C_{j_i}

and

$$B = \sum_{i=0}^{K} \binom{K}{i} \frac{(\alpha + i - 1)!}{(i - 1)!}$$

$$D = \prod_{j=1}^{c} (n_j!).$$

Then, by formulae (2.6'), (2.7'), and (2.8'),

$$P(e/C_{j_i}) = \frac{(w_i - 1)!}{(n + w_i - 1)!} D$$

$$P(C_{j_i}) = \frac{1}{B} \frac{(\alpha + w_i - 1)!}{(w_i - 1)!}$$

$$P(e) = \frac{D}{B} \sum_{i=0}^{K-c} \binom{K - c}{i} \frac{(\alpha + c + i - 1)!}{(n + c + i - 1)!}.$$

Now, evidence e states that certain c Ct-predicates are exemplified, and generalization g states that certain other b Ct-predicates are empty. Each constituent C_{j_i}, being compatible with e and implying g, says that either $c, c+1, \ldots,$ or $c+(K-b-c)=K-b$ Ct-predicates are instantiated. Therefore, $c \leqslant w_i \leqslant K-b$, for all $i = j_1, j_2, \ldots, j_g$. Moreover, for all $m = 0, 1, \cdots,$ $K-b-c$, there are $\binom{K-b-c}{m}$ constituents C_{j_i} such that $w_i = c + m$. If such a constituent is denoted by C_{c+m}, then by formula (3.1)

$$P(g/e) = \frac{1}{P(e)} \sum_{m=0}^{K-b-c} \binom{K - b - c}{m} P(C_{c+m}) P(e/C_{c+m}).$$

By combining the above formulae, we have the result

$$(3.2) \qquad P(g/e) = \frac{\sum_{i=0}^{K-b-c} \binom{K - b - c}{i} \dfrac{(\alpha + c + i - 1)!}{(n + c + i - 1)!}}{\sum_{i=0}^{K-c} \binom{K - c}{i} \dfrac{(\alpha + c + i - 1)!}{(n + c + i - 1)!}}.$$

2. EVIDENTIAL THEORETICAL CONCEPTS

2.1. Let P be a new monadic predicate which is joined to the language $L(\lambda)$, and let $L(\lambda \cup \{P\})$, or L for short, be the extended language with the $k+1$ primitive predicates $O_1, O_2, ..., O_k$, and P. The Ct-predicates of L will be denoted by $Ct_1^r, Ct_2^r, ..., Ct_{K'}^r$, where $K' = 2^{k+1} = 2K$. Each Ct-predicate Ct_i of $L(\lambda)$ is now equivalent to the disjunction of two Ct^r-predicates of L, that is to say,

$$Ct_i = Ct_{i1}^r \vee Ct_{i2}^r$$

where

$$Ct_{i1}^r = Ct_i \,\&\, P$$
$$Ct_{i2}^r = Ct_i \,\&\, {\sim} P.$$

By means of the extended language L, we can make a finer partition of the universe U into 'cells' determined by the Ct^r-predicates. The new predicate P may be connected with the old predicates $O_1, O_2, ..., O_k$ in various ways. For example, P may be explicitly, piecewise, or partially definable by $O_1, O_2, ..., O_k$ (cf. Chapter 5), or some general sentences of L containing P and some of $O_1, O_2, ..., O_k$ may be true (or, assumed to be true) by our conceptual or factual knowledge. Below, we let T be the conjunction of all these definitions or sentences. T is, then, a 'theory' expressing the connections of P to our old vocabulary. The effect of T is usually, but not always, that some cells Ct^r are empty by it, in the sense that Ct^r cannot be instantiated if T is true. In any case, it is convenient first to consider theories T which rule out some cells Ct^r of L. (The calculation of probabilities in the case of theories without this effect, e.g., piecewise definitions in Chapter 5, can generally be reduced to this case.) For simplicity, we have restricted ourselves to theories which do not claim that some kinds of individuals are instantiated in the universe. We let

$r =$ the number of Ct^r-predicates of L which are empty by T,

and assume that $r > 0$. However, it is not assumed that T has to rule out any cells Ct of $L(\lambda)$. This will happen, of course, only if T has observational consequences, i.e., non-tautological theorems in $L(\lambda)$.

We have assumed that the individuals in sample e have been observed completely with respect to the language $L(\lambda)$. In the richer language L, it may be possible to *reinterpret* this evidence in the light of the new predicate P and the theory T. In general, for each Ct-predicate $Ct_i = Ct_{i1}^r \vee \vee Ct_{i2}^r$ of $L(\lambda)$, which is exemplified in e, there are the following possibilities:

(1) (a) Ct_{i1}^r is empty by T, or

    (b) Ct_{i2}^r is empty by T.

(Both of them cannot be empty by T because T is assumed to be compatible with e.) If neither of the cases (1)(a) and (1)(b) obtains, we still have the following possibilities:

(2) The predicate P is *evidential* in the sense that it can be used for the direct reporting of our evidence. In other words, P is a new predicate which is observational with respect to the theory T or can be measured validly relying on the truth of T. Then we can 'observe' the individuals in e completely with respect to L, that is, we can state for each individual in our sample whether it satisfies P or $\sim P$. Then it may happen that

(a) Ct_{i1}^r is empty, Ct_{i2}^r is not empty, or

(b) Ct_{i1}^r is not empty, Ct_{i2}^r is empty, or

(c)   neither Ct_{i1}^r nor Ct_{i2}^r is empty.

(3) The predicate P is *non-evidential* in the sense that we cannot tell, even if we rely on the truth of the theory T, whether an individual in our

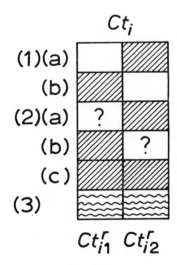

Fig. 3.1.

sample satisfies P or $\sim P$. Then we cannot determine whether an individual satisfying Ct_i belongs to Ct_{i1}^r or to Ct_{i2}^r, and our evidence e is incomplete with respect to the language L.

In this work, we make a simplifying assumption that an evidential theoretical predicate can always be used for the reporting of evidence and that a non-evidential theoretical predicate can never be used for that purpose. However, the cases between these two extremes can, in principle, be handled within our framework without any new problems other than an increase in the complexity of our calculations.

In what follows, the above cases (1)–(3) in regard to the cells of L will be schematically represented as indicated in Figure 3.1 (see p. 33). (Note the difference between cases (1)(a) and (2)(a): in (1)(a) the cell Ct_{i1}^r is known to be empty by theory T while in (2)(a) it is not known whether Ct_{i1}^r is exemplified in our universe or not.)

We shall first assume that the predicate P is evidential in sense (2), i.e., that no cases of (3) occur when the evidence is described in L. The following notations (besides T, r and g) will be used:

c = the number of Ct-predicates of $L(\lambda)$ exemplified in evidence e $(1 \leqslant c \leqslant K)$

c_o = the number of Ct-predicates of $L(\lambda)$ corresponding to the case (2)(c) above, i.e., the number of cells of $L(\lambda)$ in which the individuals of e split into both smaller cells of L $(0 \leqslant c_o \leqslant c)$

$c' = c + c_o$ = the number of cells of L exemplified by the individuals of e $(1 \leqslant c' \leqslant 2c)$

b = the number of Ct-predicates of $L(\lambda)$ which are empty by generalization g $(0 < b \leqslant K - c)$

b' = the number of Ct^r-predicates of L which are empty by g but not by T $(0 \leqslant b' \leqslant 2b)$.

The situation may be schematically represented by Figure 3.2 (see p. 35). In Figure 3.2, the small squares represent the cells Ct^r of L, and any pair of two adjacent cells (starting from the first or last column) represents a cell Ct of $L(\lambda)$. The shaded cells are those exemplified in evidence e, the white cells are those empty by theory T, and the cells with the question mark are those that are neither exemplified in e nor empty by T.

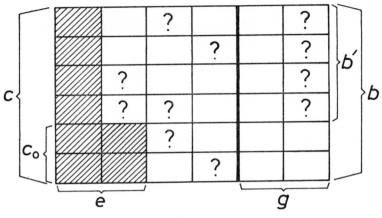

Fig. 3.2.

2.2.   We shall now compute the probability $P(g/e \ \& \ T)$ of generalization g on evidence e and theory T. Suppose that

$$e \ \& \ T \vdash g \equiv C^r_{j_1} \vee C^r_{j_2} \vee \cdots \vee C^r_{j_g}$$
$$e \vdash T \equiv C^r_{j_1} \vee \cdots \vee C^r_{j_g} \vee C^r_{j_{g+1}} \vee \cdots \vee C^r_{j_t}.$$

Then

$$P(g/e \ \& \ T) = \sum_{i=j_1}^{j_g} P(C^r_i/e \ \& \ T)$$

$$= \frac{1}{P(T/e)} \sum_{i=j_1}^{j_g} P(C^r_i \ \& \ T/e)$$

$$= \frac{1}{P(T/e)} \sum_{i=j_1}^{j_g} P(C^r_i/e)$$

$$= \frac{\sum\limits_{i=j_1}^{j_g} P(C^r_i/e)}{\sum\limits_{i=j_1}^{j_t} P(C^r_i/e)}$$

$$= \frac{\sum\limits_{i=j_1}^{j_g} P(C^r_i) \, P(e/C^r_i)}{\sum\limits_{i=j_1}^{j_t} P(C^r_i) \, P(e/C^r_i)}.$$

Let

$w_i^r =$ the number of Ct^r-predicates of L which are instantiated by the constituent C_i^r.

Then by formulae (2.6′) and (2.7′),

$$P(C_i^r)\, P(e/C_i^r) = M \,\frac{(\alpha + w_i^r - 1)!}{(n + w_i^r - 1)!}$$

where M is a constant depending on K and the number of individuals of e in the cells of L (but not on i). Hence

$$P(g/e\ \&\ T) = \frac{\displaystyle\sum_{i=j_1}^{j_g} \frac{(\alpha + w_i^r - 1)!}{(n + w_i^r - 1)!}}{\displaystyle\sum_{i=j_1}^{j_t} \frac{(\alpha + w_i^r - 1)!}{(n + w_i^r - 1)!}}\ .$$

According to e, at least c' cells of L are exemplified. According to T, at least r cells of L are not instantiated in the universe. Hence, $c' \leqslant w_i^r \leqslant 2K - r$, for all $i = j_1, \ldots, j_g, \ldots, j_t$. According to g, there are b' empty cells in addition to those which are empty by T. Hence, $c' \leqslant w_i^r \leqslant 2K - b' - r$, for all $i = j_1, \ldots, j_g$. Moreover, among the constituents C_i^r, $i = j_1, \ldots, j_g, \ldots, j_t$, there are $\binom{2K-r-c'}{m}$ such that $w_i^r = c' + m$, where $m = 0, 1, \ldots, 2K - r - c'$. Similarly, among the constituents C_i^r, $i = j_1, \ldots, j_g$, there are $\binom{2K-r-b'-c'}{m}$ such that $w_i^r = c' + m$, where $m = 0, 1, \ldots, 2K - r - b' - c'$. (Note that $2K - r - c'$ is the number of question marks in Figure 3.2, and that $2K - r - b' - c'$ is the number of cells with a question mark outside those cells that are empty by g.) Thus, we get the result

$$(3.3)\quad P(g/e\ \&\ T) = \frac{\displaystyle\sum_{i=0}^{2K-r-b'-c'} \binom{2K - r - b' - c'}{i} \frac{(\alpha + c' + i - 1)!}{(n + c' + i - 1)!}}{\displaystyle\sum_{i=0}^{2K-r-c'} \binom{2K - r - c'}{i} \frac{(\alpha + c' + i - 1)!}{(n + c' + i - 1)!}}\ .$$

The formulae (3.2) and (3.3) give us the probabilities $P(g/e)$ in $L(\lambda)$ and $P(g/e\ \&\ T)$ in L. If we put $n = 0$ and $c = 0$ in formula (3.2), we get the *a priori* probability of generalization g in $L(\lambda)$:

$$(3.4) \qquad P(g) = \frac{\sum_{i=0}^{K-b} \binom{K-b}{i} \frac{(\alpha+i-1)!}{(i-1)!}}{\sum_{i=0}^{K} \binom{K}{i} \frac{(\alpha+i-1)!}{(i-1)!}}.$$

If we put $r=0$ and $b'=2b$ in formula (3.3), we get the probability of generalization g on evidence e in the richer language L without any theoretical knowledge about the new predicate P. Similarly, if we put $r=0$, $b'=2b$, $c'=0$, and $n=0$ in (3.3), we get the *a priori* probability of g in L. Finally, the substitution of $c'=0$ and $n=0$ into (3.3) gives us the probability of g on theory T in L, without any observational evidence:

$$(3.5) \qquad P(g/T) = \frac{\sum_{i=0}^{2K-r-b'} \binom{2K-r-b'}{i} \frac{(\alpha+i-1)!}{(i-1)!}}{\sum_{i=0}^{2K-r} \binom{2K-r}{i} \frac{(\alpha+i-1)!}{(i-1)!}}.$$

If $b'=0$, then $(T \& e) \vdash g$ and $P(g/e \& T)=1$, i.e., we have a case of deductive systematization. Therefore, we shall always assume that $b'>0$.

We shall not here study formulae (3.2) and (3.3) in their general form. Instead, we investigate their behaviour in some interesting special cases.

(1) If $n \to \infty$ and $\alpha \neq \infty$, then $P(g/e)$ and $P(g/e \& T)$ approach one.

(2) If $\alpha \to \infty$, then $P(g/e)$ and $P(g/e \& T)$ approach zero.

(3) If $r=K$, $c'=c$, and $b'=b$, then $P(g/e)=P(g/e \& T)$. An example of this case is the situation where P is *explicitly definable* in terms of O_1, $O_2,..., O_k$, and T is this definition. In fact, in this case the individuals in a cell Ct_i of $L(\lambda)$ either all have P or all have $\sim P$, i.e., in no cell of $L(\lambda)$ may the individuals split into both smaller cells.

(4) If $\alpha=n$, then

$$P(g/e) = \frac{\sum_{i=0}^{K-b-c} \binom{K-b-c}{i}}{\sum_{i=0}^{K-c} \binom{K-c}{i}} = \frac{2^{K-b-c}}{2^{K-c}} = 2^{-b}$$

and similarly,

$$P(g/e \& T) = 2^{-b'}.$$

In this case,

$$P(g/e) < P(g/e \,\&\, T) \quad \text{iff} \quad b > b'.$$

(5) If n and α are sufficiently large in relation to $(K-1)^2$ and $(2K-r-1)^2$, then – using the approximation $(m+n)! \simeq m! m^n$, where m is sufficiently large in relation to n^2 (cf. Carnap, 1950, p. 150) – we get

$$P(g/e) \simeq \frac{\sum\limits_{i=0}^{K-b-c} \binom{K-b-c}{i}\left(\dfrac{\alpha}{n}\right)^i}{\sum\limits_{i=0}^{K-c} \binom{K-c}{i}\left(\dfrac{\alpha}{n}\right)^i} = \frac{\left(1+\dfrac{\alpha}{n}\right)^{K-c-b}}{\left(1+\dfrac{\alpha}{n}\right)^{K-c}} \cdot$$

Hence, asymptotically (for great n and α)

(3.6) $\quad P(g/e) \simeq \left(1 + \dfrac{\alpha}{n}\right)^{-b}$

and similarly

(3.7) $\quad P(g/e \,\&\, T) \simeq \left(1 + \dfrac{\alpha}{n}\right)^{-b'}$

(3.8) $\quad P(g) \simeq (1 + \alpha)^{-b}.$

Also in this case we have the result

(3.9) $\quad P(g/e) < P(g/e \,\&\, T) \quad \text{iff} \quad b > b'.$

Note that $b > b'$ only if the theory T has observational consequences. Therefore, we have asymptotically

(3.10) $\quad P(g/e) < P(g/e \,\&\, T)$ only if T has observational consequences.

This shows that, asymptotically, we can increase the probability of an observational generalization g by introducing a new predicate P into our language only if we have at our disposal a fairly strong theory on the relations of this new predicate to the old vocabulary. In other words, if we have a large sample of individuals from a fairly irregular universe, then the redescription of the universe by means of a richer vocabulary increases the probability of a generalization in the old vocabulary only if the new predicate is fairly strongly connected or involved with the old vocabulary.

From formulae (3.6) and (3.8) we see that asymptotically

$$P(g) < P(g/e).$$

However, it is possible that the probability $P(g/e \ \& \ T)$ is asymptotically less than the *a priori* probability of g if $b' > b$. By formulae (3.7) and (3.8)

(3.11) $P(g) > P(g/e \ \& \ T)$ iff $\left(1 + \dfrac{\alpha}{n}\right)^{b'} > (1 + \alpha)^b.$[1]

In particular, we have

(3.12)   If $b' = 2b$, then

$$P(g) > P(g/e \ \& \ T) \quad \text{iff} \quad \alpha > n(n - 2).$$

(6) The case where $b + c = K$ is of special interest. In this case, the generalization g states that the universe contains only individuals of the kind already exemplified in the evidence. Formula (3.2) is then reduced to

$$P(g/e) = \frac{\dfrac{(\alpha + c - 1)!}{(n + c - 1)!}}{\displaystyle\sum_{i=0}^{b} \binom{b}{i} \dfrac{(\alpha + c + i - 1)!}{(n + c + i - 1)!}}$$

or

(3.13) $P(g/e) = \dfrac{1}{1 + A},$

where

$$A = \frac{\alpha + c}{n + c} \times$$

$$\times \left\{ \binom{b}{1} + \binom{b}{2}\frac{(\alpha + c + 1)}{(n + c + 1)} + \cdots + \binom{b}{b}\frac{(\alpha + c + 1)\ldots(\alpha + c + b - 1)}{(n + c + 1)\ldots(n + c + b - 1)} \right\}.$$

Similarly, if we assume that $b' + c' + r = 2K$ (i.e., that theory T leaves no question marks outside those cells which are empty by g), then formula (3.3) reduces to

(3.14) $P(g/e \ \& \ T) = \dfrac{1}{1 + B},$

where

$$B = \frac{\alpha + c'}{n + c'} \left\{ \binom{b'}{1} + \binom{b'}{2} \frac{(\alpha + c' + 1)}{(n + c' + 1)} + \right.$$

$$\left. + \cdots + \binom{b'}{b'} \frac{(\alpha + c' + 1) \ldots (\alpha + c' + b' - 1)}{(n + c' + 1) \ldots (n + c' + b' - 1)} \right\}.$$

In particular, if $b = 1$ and $b' = 1$, then

$$P(g/e) = \frac{1}{1 + \dfrac{\alpha + c}{n + c}}$$

and

$$P(g/e \ \& \ T) = \frac{1}{1 + \dfrac{\alpha + c + c_o}{n + c + c_o}}.$$

In this case, we have

$$P(g/e) < P(g/e \ \& \ T) \quad \text{iff} \quad \frac{\alpha + c}{n + c} > \frac{\alpha + c + c_o}{n + c + c_o}$$

$$\text{iff} \quad \alpha > n \quad \text{and} \quad c_o > 0.$$

The same result holds more generally:

(3.15)   If $b + c = K$, $b' + c' + r = 2K$, and $b = b'$, then
$P(g/e) < P(g/e \ \& \ T)$ iff $\alpha > n$ and $c_o > 0$.

Note also that, by formulae (3.13) and (3.14),

$$\alpha > n \quad \text{iff} \quad P(g/e) < (\tfrac{1}{2})^b$$
$$\text{iff} \quad P(g/e \ \& \ T) < (\tfrac{1}{2})^{b'}.$$

In the case just considered, g is a generalization to the effect that the universe U has the same characteristics as our sample e, i.e., that the Ct-cells exemplified in e are instantiated in U and all other Ct-cells are empty. By introducing the new predicate P, we split these Ct-cells into two parts. T is a theory which implies that half of the smaller Ct^r-cells, which result from splitting the Ct-cells empty by g, are empty. Moreover, T implies that those Ct^r-cells, which result from splitting the Ct-cells exemplified in e but fail to be exemplified, are empty. In this situation, the probability of g may increase by T if and only if two conditions (cf. result (3.15)) are satisfied. In the first place, evidence e must show a kind of variety (or openness) with respect to the richer language. Namely, individuals in at least one Ct-cell exemplified in e have to split into both subcells (with respect to P and $\sim P$). In the second place, the number of observed individuals must not exceed the value of the parameter α. This means that evidence e is fairly weak with respect to g in the sense that the probability of g is small (at least less than $\frac{1}{2}$). Moreover, even if $P(g/e \& T)$ is here greater than $P(g/e)$, it is still less than $\frac{1}{2}$.

2.3.   This section will be concluded by brief consideration of the more general case in which the richer language L contains $m > 1$ evidential theoretical predicates. As we shall see, there are no essential differences between this case and the case of only one theoretical predicate. Since we have qualitatively the same results in both cases, we shall elsewhere in this work consider only the case $\mu = \{P\}$.

Suppose that $\lambda = \{O_1, ..., O_k\}$ and $\mu = \{P_1, ..., P_m\}$. Then each Ct-predicate of $L(\lambda)$ splits into 2^m Ct^r-predicates of $L = L(\lambda \cup \mu)$. Thus, the language L has $2^m K$ Ct^r-predicates (where K is 2^k, as earlier). We let T, g, e, b, b', r, c, c', and c_o have the same meaning as above. Here

$$0 \leqslant b' \leqslant 2^m b$$

and

$$1 \leqslant c' \leqslant 2^m c.$$

Corresponding to (3.3), the probability of a weak generalization g of $L(\lambda)$, given the evidence e and the theory T in L, can now be computed from the formula:

(3.3') $P(g/e \& T) =$

$$= \frac{\sum\limits_{i=0}^{2^m K - r - b' - c'} \binom{2^m K - r - b' - c'}{i} \dfrac{(\alpha + c' + i - 1)!}{(n + c' + i - 1)!}}{\sum\limits_{i=0}^{2^m K - r - c'} \binom{2^m K - r - c'}{i} \dfrac{(\alpha + c' + i - 1)!}{(n + c' + i - 1)!}}.$$

(3.3') differs from (3.3) only in that $2K$ has been replaced by $2^m K$. In fact, (3.3) is obtained from (3.3') as a special case by putting $m = 1$.

Exactly the same conclusions as (1)–(6) above follow from (3.3'). (The only differences are that in (3) we have now $r=2^{m-1}K$ and in (6) $b'+c'+ +r=2^mK$.)

To illustrate formula (3.3'), let $\lambda=\{O_1, O_2\}$ and $\mu=\{P_1, P_2\}$, and let T be a theory in L corresponding to Hempel's theory about white phosphorus (cf. Chapter 1.1), i.e.,

$$T = (x) \, [(P_1(x) \supset O_1(x)) \, \& \, (P_1(x) \supset P_2(x)) \, \& \\ \& \, (P_2(x) \supset O_2(x))].$$

Then $k=2, K=4, m=2$, and $r=9$. Suppose that the Ct-predicates $O_1 \& O_2$, $\sim O_1 \& O_2$, and $\sim O_1 \& \sim O_2$ have been exemplified in the evidence e, and that the individuals of e split into all the subcells that are not empty by T. Then $c=3$ and $c'=6$. Let g be the generalization

$$g = (x) \, (O_1(x) \supset O_2(x)),$$

so that $b=1$ and $b'=1$. The situation can be represented by the following figure:

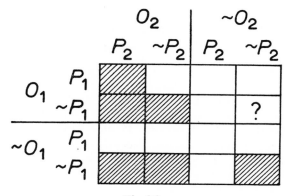

Fig. 3.3.

Now we have, by (3.2) and (3.3'),

$$P(g/e) = \cfrac{1}{1 + \cfrac{\alpha + 3}{n + 3}}$$

and

$$P(g/e \ \& \ T) = \frac{1}{1 + \dfrac{\alpha + 6}{n + 6}}.$$

Hence,

$$P(g/e) < P(g/e \ \& \ T) \quad \text{iff} \quad n < \alpha.$$

3. Non-evidential theoretical concepts

We have assumed above that P is an evidential predicate in the sense that it can be used for the direct reporting of evidence. We now consider the case in which P is a *non-evidential theoretical* predicate so that for no individual in our evidence e can we say whether it has the property P or $\sim P$ – unless theory T excludes one of these two possibilities. This case can, in principle, be handled in the same way as the earlier one, even if the incompleteness of our evidence complicates the situation.

Let $Ct_{i_1}, Ct_{i_2}, \ldots, Ct_{i_c}$ be the Ct-predicates which are exemplified in the sample e, and let $n_1, n_2, \ldots, n_c$, respectively, be the number of individuals in these cells. Let $Ct_{i_1}, Ct_{i_2}, \ldots, Ct_{i_d}$ be those Ct-predicates for which neither $Ct_{i_j} \ \& \ P$ nor $Ct_{i_j} \ \& \sim P$ is empty by T ($j = 1, \ldots, d$). Then, of course, $Ct_{i_{d+1}}, Ct_{i_{d+2}}, \ldots, Ct_{i_c}$ are those Ct-predicates for which either

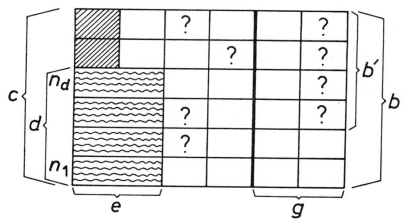

Fig. 3.4.

Ct_{i_j} & P or Ct_{i_j} & $\sim P$ (but not both) is empty by T ($j=d+1,...,c$). The situation may be schematically represented by Figure 3.4 (see p. 43). There are, again, $2K$ cells r of which are empty by T and b' empty by generalization g but not by T. The total number of question marks is $2K-r-c-d$. Let

$$e \vdash T \equiv C^r_{j_1} \vee \cdots \vee C^r_{j_g} \vee \cdots \vee C^r_{j_t}$$
$$e \mathbin{\&} T \vdash g \equiv C^r_{j_1} \vee \cdots \vee C^r_{j_g}.$$

Let C^r_i be a constituent which is compatible with e. Then, for each Ct_{i_j}, $j=1,...,d$, we have three possibilities:

1. Ct_{i_j} & P is empty, or
2. Ct_{i_j} & $\sim P$ is empty, or
3. Neither Ct_{i_j} & P nor Ct_{i_j} & $\sim P$ is empty

according to the constituent C^r_i. For each C^r_i, $i=j_1,...,j_t$, let

$d'_i=$ the number of Ct-predicates Ct_{i_j}, $j=1,...,d$, which are of type 3 according to C^r_i.

Then $0 \leqslant d'_i \leqslant d$. For $i=j_1,...,j_t$, let

$w^r_i=$ the number of Ct^r-predicates instantiated according to C^r_i.

Then

$$c + d'_i \leqslant w^r_i \leqslant 2K - r, \quad \text{for} \quad i = j_1, ..., j_t$$
$$c + d'_i \leqslant w^r_i \leqslant 2K - r - b', \quad \text{for} \quad i = j_1, ..., j_g.$$

For $i=j_1,...,j_t$, let

$$n^i_j = n_j + 1, \quad \text{if} \quad Ct_{i_j} \text{ is of type 3 according to } C^r_i$$
$$= n_j, \quad \text{otherwise}$$

($j=1,...,d$).

Now, by the formula (2.10) we get

$$P(e/C^r_i) = \frac{(w^r_i - 1)!}{(n + w^r_i - 1)!} \, n^i_1! n^i_2! \ldots n^i_d! n_{d+1}! \ldots n_c!$$

and

$$P(C_i^r)\,P(e/C_i^r) = \frac{(\alpha + w_i^r - 1)!}{(n + w_i^r - 1)!}\, n_1^i!\,n_2^i!\,...\,n_d^i!M,$$

where M is a constant depending on K, α, n_{d+1}, n_{d+2}, ..., n_c, but not on i. Hence

(3.16) $$P(g/e\ \&\ T) = \frac{\displaystyle\sum_{i=j_1}^{j_g} \frac{(\alpha + w_i^r - 1)!}{(n + w_i^r - 1)!}\, n_1^i!\,n_2^i!\,...\,n_d^i!}{\displaystyle\sum_{i=j_1}^{j_t} \frac{(\alpha + w_i^r - 1)!}{(n + w_i^r - 1)!}\, n_1^i!\,n_2^i!\,...\,n_d^i!}.$$

In general, for each $d_i'=0, 1, ..., d$, there are $\binom{d}{d_i'}$ ways of choosing those d_i' Ct-predicates $Ct_{i,j}$ which are of type 3 according to C_i^r. For each choice of this kind, there are $2^{d-d_i'}\binom{2K-r-c-d}{m}$ constituents C_i^r, $i=j_1, ..., j_t$, such that $w_i^r=c+d_i'+m$, where $m=0, 1, ..., 2K-r-c-d$. And for each choice, there are $2^{d-d_i'}\binom{2K-r-c-d-b'}{m}$ constituents C_r^i, $i=j_1, ..., j_g$, such that $w_i^r=c+d_i'+m$, where $m=0, 1, ..., 2K-r-c--d-b'$. Moreover, for each choice of this kind, the product $n_1^i!\,n_2^i!...n_d^i!$ is different.

To simplify the calculations, let us assume that $d=1$. Then d_i' can be either 0 or 1. For a constituent C_i^r with $d_i'=0$, $n_1^i=n_1$. Now, among the constituents C_i^r, $i=j_1, ..., j_t$, there are $2\binom{2K-r-c-1}{m}$ such constituents that $d_i'=0$ and $w_i^r=c+m$, for each $m=0, 1, ..., 2K-r-c-1$. Among C_i^r, $i=j_1, ..., j_g$, there are $2\binom{2K-r-c-b'-1}{m}$ such constituents that $d_i'=0$ and $w_i^r=c+m$, for each $m=0, 1, ..., 2K-r-c-b'-1$. For a constituent C_i^r with $d_i'=1$, $n_1^i=n_1+1$. Now, among the constituents C_i^r, $i=j_1, ..., j_t$, there are $\binom{2K-r-c-1}{m}$ such constituents that $d_i'=1$ and $w_i^r=c+1+m$, where $m=0, 1, ... 2K-r-c-1$. Among C_i^r, $i=j_1, ..., j_g$, there are $\binom{2K-r-c-b'-1}{m}$ such constituents that $d_i'=1$ and $w_i^r=c+1+m$, where $m=0, 1, ..., 2K-r-c-b'-1$. Hence, by (3.16)

(3.17) $\quad P(g/e \& T) =$

$$\frac{\displaystyle\sum_{m=0}^{2K-r-c-b'-1} \binom{2K-r-c-b'-1}{m} \times \times \left\{ 2\frac{(\alpha+c+m-1)!}{(n+c+m-1)!} + (n_1+1)\frac{(\alpha+c+m)!}{(n+c+m)!} \right\}}{\displaystyle\sum_{m=0}^{2K-r-c-1} \binom{2K-r-c-1}{m} \times \times \left\{ 2\frac{(\alpha+c+m-1)!}{(n+c+m-1)!} + (n_1+1)\frac{(\alpha+c+m)!}{(n+c+m)!} \right\}}$$

Formula (3.17) can be written in the form

(3.18) $\quad P(g/e \& T) =$

$$\frac{\displaystyle\sum_{m=0}^{2K-r-c-b'-1} \binom{2K-r-c-b'-1}{m} v(m)\frac{(\alpha+c+m)!}{(n+c+m)!}}{\displaystyle\sum_{m=0}^{2K-r-c-1} \binom{2K-r-c-1}{m} v(m)\frac{(\alpha+c+m)!}{(n+c+m)!}}$$

where

$$v(m) = 2\frac{n+c+m}{\alpha+c+m} + (n_1+1).$$

From formula (3.17), we see that $P(g/e \& T)$ approaches one if $n \to \infty$ and $\alpha \neq \infty$, and that $P(g/e \& T)$ approaches zero if $\alpha \to \infty$.

If $\alpha = n$, formula (3.17) is reduced to

$$P(g/e \& T) = 2^{-b'},$$

as in the case where the predicate P is evidential.

If α and n are sufficiently great, then (letting h stand for $2K-r-c-1$)

$P(g/e \& T) \simeq$

$$\simeq \frac{\displaystyle 2\sum_{m=0}^{h-b'} \binom{h-b'}{m}\left(\frac{\alpha}{n}\right)^{m-1} + (n_1+1)\sum_{m=0}^{h-b'} \binom{h-b'}{m}\left(\frac{\alpha}{n}\right)^{m}}{\displaystyle 2\sum_{m=0}^{h} \binom{h}{m}\left(\frac{\alpha}{n}\right)^{m-1} + (n_1+1)\sum_{m=0}^{h} \binom{h}{m}\left(\frac{\alpha}{n}\right)^{m}}$$

$$= \frac{\left(2\dfrac{n}{\alpha} + n_1 + 1\right) \sum\limits_{m=0}^{h-b'} \dbinom{h-b'}{m} \left(\dfrac{\alpha}{n}\right)^m}{\left(2\dfrac{n}{\alpha} + n_1 + 1\right) \sum\limits_{m=0}^{h} \dbinom{h}{m} \left(\dfrac{\alpha}{n}\right)^m}$$

$$= \left(1 + \frac{\alpha}{n}\right)^{-b'}.$$

This is again the same result as in the case where P is evidential. In other words, for great α and n, the nature of the new predicate P does not effect a difference to the probabilities of observational generalizations.

Due to our simplifying assumption $d=1$, formula (3.18) corresponds to the earlier formula (3.3) with $c'=c+1$. The only difference between these formulae is the factor $v(m)$ in (3.18). This difference can be illustrated by a simple example.

Example 3.1. Let $L(\lambda)$ have two primitive predicates O_1 and O_2, so that $k=2$ and $K=4$. Then the Ct-predicates of $L(\lambda)$ are

$$Ct_1 = O_1 \,\&\, O_2$$
$$Ct_2 = O_1 \,\&\, \sim O_2$$
$$Ct_3 = \sim O_1 \,\&\, O_2$$
$$Ct_4 = \sim O_1 \,\&\, \sim O_2.$$

Let g be the generalization

$$g = (x)\,(O_1(x) \supset O_2(x)),$$

which says, in effect, that cell Ct_2 is empty. Hence, $b=1$. The richer language L with the new predicate P has eight Ct^r-predicates:

$$Ct_1^r = O_1 \,\&\, O_2 \,\&\, P$$
$$Ct_2^r = O_1 \,\&\, O_2 \,\&\, \sim P$$
$$Ct_3^r = O_1 \,\&\, \sim O_2 \,\&\, P$$
$$Ct_4^r = O_1 \,\&\, \sim O_2 \,\&\, \sim P$$
$$Ct_5^r = \sim O_1 \,\&\, O_2 \,\&\, P$$
$$Ct_6^r = \sim O_1 \,\&\, O_2 \,\&\, \sim P$$
$$Ct_7^r = \sim O_1 \,\&\, \sim O_2 \,\&\, P$$
$$Ct_8^r = \sim O_1 \,\&\, \sim O_2 \,\&\, \sim P.$$

Now g says that cells Ct_3^r and Ct_4^r are empty. Let T be the theory

$$T = (x) [(P(x) \supset O_1(x)) \mathbin{\&} (P(x) \supset O_2(x))].$$

Then T says that cells Ct_3^r, Ct_5^r, and Ct_7^r are empty.

Suppose now that the Ct-predicates Ct_1, Ct_3, and Ct_4 are exemplified in evidence e. Suppose also that the predicate P is evidential and that the individuals of e in cell Ct_1 split into both Ct_1^r and Ct_2^r. This situation can be represented by the following figure:

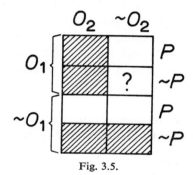

Fig. 3.5.

Here $c'=4$, $r=3$, and $b'=1$. Then, by formulae (3.13) and (3.14),

$$P(g/e) = \cfrac{1}{1 + \cfrac{\alpha + 3}{n + 3}} \quad (= p_1, \text{ see below})$$

$$P(g/e \mathbin{\&} T) = \cfrac{1}{1 + \cfrac{\alpha + 4}{n + 4}} \quad (= p_3).$$

Suppose then that all sampled individuals in cell Ct_1 happen to belong to cell Ct_1^r. We then have the situation represented by Figure 3.6. In this case, $c'=3$, $r=3$, and $b'=1$. Hence, $2K - b' - c' - r = 1$. The probability $P(g/e \mathbin{\&} T)$ is now by formula (3.3)

$$P(g/e \mathbin{\&} T) = \cfrac{1}{1 + D\,\cfrac{\alpha + 3}{n + 3}} \quad (= p_2)$$

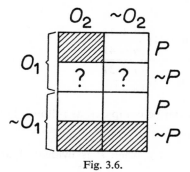

Fig. 3.6.

where

$$D = \frac{1 + \dfrac{\alpha + 4}{n + 4}}{1 + \dfrac{\alpha + 3}{n + 3}}.$$

Suppose then that the predicate P is non-evidential, i.e., that we have the following situation:

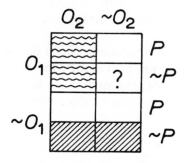

Fig. 3.7.

Now $2K - r - c - 1 = 1$ and $2K - r - c - b' - 1 = 0$. If n_1 is the number of sampled individuals in cell Ct_1, then formula (3.18) reduces to

$$P(g/e \text{ \& } T) = \frac{1}{1 + A\dfrac{\alpha + 3}{n + 3}} \quad (= p_4)$$

where

$$A = \frac{2 + (n_1 + 1)\dfrac{(\alpha + 4)}{(n + 4)}}{2 + (n_1 + 1)\dfrac{(\alpha + 3)}{(n + 3)}}.$$

These probabilities can be compared as follows. Let

$p_1 = P\,(g/e)$

$p_2 = P\,(g/e\ \&\ T)$, when P is evidential and Ct_1 is of type (2) (a) or (2) (b) (cf. p. 33)

$p_3 = P\,(g/e\ \&\ T)$, when P is evidential and Ct_1 is of type (2) (c)

$p_4 = P\,(g/e\ \&\ T)$, when P is non-evidential.

Then, in these examples,

If $\alpha > n$, then $p_1 < p_2 < p_4 < p_3 < \tfrac{1}{2}$.

If $\alpha < n$, then $p_1 > p_2 > p_4 > p_3 > \tfrac{1}{2}$.

If $\alpha = n$, then $p_1 = p_2 = p_4 = p_3 = \tfrac{1}{2}$.

As we see from this result, the case of an evidence incomplete with respect to L (P non-evidential) falls here between the cases of complete evidence (P evidential) of type (a) (or (b)) and of type (c).

To conclude this chapter, it should be noted that we have now defined probability measures P_0 for $L(\lambda)$ and P_1 for L so that our adequacy conditions CA2, CA3, CA6, and partly CA1 are satisfied. The next two chapters contain further technical considerations of measures P_0 and P_1.

NOTE

[1] The customary abbreviation 'iff' for 'if and only if' is used here.

CHAPTER 4

INDUCTIVE PROBABILITIES
OF STRONG GENERALIZATIONS

In Chapter 3, we considered weak generalizations of $L(\lambda)$, i.e., disjunctions of constituents to the effect that *at least* certain b Ct-predicates of $L(\lambda)$ are empty. In this chapter, a study is made of the probabilities of strong generalizations, i.e., constituents of the observational language. (The formulas of Chapter 3 cannot, in general, be applied to this case.)

Let g be a constituent C_w of $L(\lambda)$ which is compatible with the evidence e, according to which c Ct-predicates of $L(\lambda)$ are exemplified. If $w = {} = K - b$, then g is a strong generalization according to which the Ct-predicates $Ct_{i_1}, Ct_{i_2}, ..., Ct_{i_w}$ are instantiated in the universe and the other b Ct-predicates are empty. In the richer language L (where the new predicate P is assumed to be evidential) there are 3^w constituents C_i^r compatible with C_w. For each $Ct_j, j = i_1, i_2, ..., i_w$, there are three possibilities: either (i) $Ct_j \& P$, or (ii) $Ct_j \& {\sim}P$, or (iii) both of them are instantiated in the universe according to a constituent C_i^r. Let T be a theory in L, which is compatible with g, and let

$$e \mathbin{\&} T \vdash {} \equiv C_{j_1}^r \lor C_{j_2}^r \lor \cdots \lor C_{j_g}^r.$$

Let $Ct_{i_{c+1}}, Ct_{i_{c+2}}, ..., Ct_{i_w}$ be the Ct-predicates instantiated by C_w but not exemplified in e. Their number is $w - c = K - b - c$. Suppose, moreover, that among them there are d such Ct-predicates that theory T allows all the cases (i)–(iii) with respect to them, i.e., that T does not imply that individuals in them must all have P or all have $\sim P$. The situation can then be represented by Figure 4.1. In this figure, $b, b', c, c_o, c' = c + c_o$ are as in Chapter 3, d is the number of Ct^r-cell pairs 'outside' e and g with two question marks ($0 \leqslant d \leqslant K - b - c$), and c_1 is the number of question marks 'inside' e.

For simplicity, we shall assume that $c_1 = 0$.[1] Then $c' + r + (K - b - c) + {} + d + b' = 2K$. Hence,

$$d = K - r - c_o + b - b'.$$

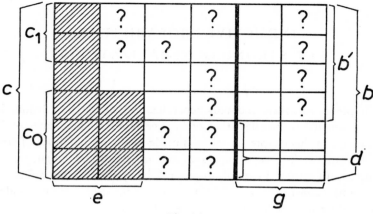

Fig. 4.1.

Let

$$q = K - b + c_o = 2K - r - d - b'.$$

Among the constituents $C^r_{j_1}, C^r_{j_2}, ..., C^r_{j_g}$, there are 2^d such that $w^r_i = q$. In general, we have

$$q \leqslant w^r_i \leqslant q + d$$

for all $i = j_1, ..., j_g$. Moreover, there are $\binom{d}{m} 2^{d-m}$ such constituents $C^r_i, i = j_1, ..., j_g$, that $w^r_i = q + m$, for $m = 0, ..., d$. In the same way as formulae (3.2)–(3.4) were obtained, we derive the results

$$(4.1) \qquad P(C_w) = \frac{\dfrac{(\alpha + w - 1)!}{(w - 1)!}}{\displaystyle\sum_{i=0}^{K} \binom{K}{i} \dfrac{(\alpha + i - 1)!}{(i - 1)!}}$$

$$(4.2) \qquad P(C_w/e) = \frac{\dfrac{(\alpha + w - 1)!}{(n + w - 1)!}}{\displaystyle\sum_{i=0}^{K-c} \binom{K - c}{i} \dfrac{(\alpha + c + i - 1)!}{(n + c + i - 1)!}}$$

where $w = K - b$, and

$$(4.3) \qquad P(C_w/e \ \& \ T) = \frac{\displaystyle\sum_{i=0}^{d} \binom{d}{i} 2^{d-i} \frac{(\alpha + q + i - 1)!}{(n + q + i - 1)!}}{\displaystyle\sum_{i=0}^{2K-c'-r} \binom{2K - c' - r}{i} \frac{(\alpha + c' + i - 1)!}{(n + c' + i - 1)!}}$$

where $q = K - b + c_o$ and $d = K - r - c_o + b - b'$.

Above we have computed the probability $P(C_w/e \ \& \ T)$ only in the case where P is a new evidential predicate. If P is nonevidential, this probability can be computed in a similar way as in deriving formulae (3.16)–(3.18).

We now study formulae (4.1)–(4.3) in some special cases.

(1) If T is an explicit definition of P in terms of $O_1, O_2, ..., O_k$, then $c_1 = 0$, $c_o = 0$, $K = r$, and $b' = b$. Hence, $c' = c$, $q = K - b$, and $d = 0$. In this case, $P(C_w/e) = P(C_w/e \ \& \ T)$.

(2) If $\alpha = n$, we have

$$P(C_w/e) = \frac{1}{2^{K-c}}$$

and

$$P(C_w/e \ \& \ T) = \frac{\displaystyle\sum_{i=0}^{d} \binom{d}{i} 2^{d-i}}{2^{2K-c'-r}} = \frac{2^d (1 + \tfrac{1}{2})^d}{2^{2K-c'-r}}$$

$$= \frac{3^d}{2^{2K-c-r}}.$$

In this case,

$$P(C_w/e) < P(C_w/e \ \& \ T) \quad \text{iff} \quad 2^{K-r-c_o} < 3^d$$

where $d = K - r - c_o + b - b'$. Hence,

(4.4) \qquad If $\alpha = n$, then
$$P(C_w/e) < P(C_w/e \ \& \ T) \quad \text{iff} \quad (\tfrac{2}{3})^{K-r-c_o} < 3^{b-b'}.$$

A sufficient condition for this is

$$K - r - c_o > 0 \quad \text{and} \quad b \geqslant b'.$$

More generally, we have (approtimately)

If $\alpha = n$, then

$$P(C_w/e) < P(C_w/e \ \& \ T) \quad \text{iff} \quad (K - r - c_o) + 2.6(b - b') > 0.$$

(3) Suppose that α and n are great. Then

$$(4.5) \qquad P(C_w/e) \simeq \frac{\left(\dfrac{\alpha}{n}\right)^{w-c}}{\displaystyle\sum_{i=0}^{K-c} \binom{K-c}{i} \left(\dfrac{\alpha}{n}\right)^{i}} = \frac{\left(\dfrac{\alpha}{n}\right)^{K-b-c}}{\left(1 + \dfrac{\alpha}{n}\right)^{K-c}}$$

and

$$P(C_w/e \ \& \ T) \simeq \frac{2^d \left(\dfrac{\alpha}{n}\right)^{q} \displaystyle\sum_{i=0}^{d} \binom{d}{i} 2^{-i} \left(\dfrac{\alpha}{n}\right)^{i}}{\left(\dfrac{\alpha}{n}\right)^{c'} 2^{K-c'-r} \displaystyle\sum_{i=0}^{2K-c'-r} \binom{2K-c'-r}{i} \left(\dfrac{\alpha}{n}\right)^{i}}$$

$$= 2^d \left(\dfrac{\alpha}{n}\right)^{q-c'} \frac{\left(1 + \dfrac{\alpha}{2n}\right)^{d}}{\left(1 + \dfrac{\alpha}{n}\right)^{2K-c'-r}}$$

$$(4.6) \qquad = \frac{\left(\dfrac{\alpha}{n}\right)^{K-b-c} \left(2 + \dfrac{\alpha}{n}\right)^{d}}{\left(1 + \dfrac{\alpha}{n}\right)^{2K-c'-r}}$$

where $d = K - r - c_o + b - b'$. If $K - b - c = 0$, then $d = 0$ and $2K - c' - r = b'$, so that formulae (4.5) and (4.6) are reduced to

$$P(C_w/e) \simeq \left(1 + \dfrac{\alpha}{n}\right)^{-b}$$

$$P(C_w/e \ \& \ T) \simeq \left(1 + \dfrac{\alpha}{n}\right)^{-b'}.$$

From formulae (4.5) and (4.6), we see that $P(C_w/e)$ and $P(C_w/e \ \& \ T)$ approach zero, if $\alpha \to \infty$. Moreover, when $n \to \infty$ and $\alpha \neq \infty$

(4.7) $P(C_w/e) \rightarrow \begin{cases} 0, \text{ if } K - b - c > 0 \\ 1, \text{ if } K - b - c = 0. \end{cases}$

$P(C_w/e \ \& \ T) \rightarrow \begin{cases} 0, \text{ if } K - b - c > 0 \\ 1, \text{ if } K - b - c = 0. \end{cases}$

By means of formulae (4.5) and (4.6), we arrive at the following result for great α and n

(4.8) $P(C_w/e) < P(C_w/e \ \& \ T)$ iff

$$\left(1 + \frac{\alpha}{n}\right)^{K-r-c_0} < \left(2 + \frac{\alpha}{n}\right)^{K-r-c_0+b-b'}.$$

A sufficient condition for this is

$$K - r - c_o > 0 \quad \text{and} \quad b \geqslant b'.$$

This can be illustrated by the following example.

Example 4.1. Let $L(\lambda)$ have two primitive predicates (i.e., $k=2$, $K=4$) and let C_w be the constituent according to which cell $O_1 \ \& \sim O_2$ is empty. Then $b=1$ and $w=3$. Suppose that cells $\sim O_1 \ \& \ O_2$ and $\sim O_1 \ \& \sim O_2$ have been exemplified in a sample e. Hence, $c=2$. Let T be the theory

$$T = (x) \ [(P \ (x) \supset O_1(x)) \ \& \ (P \ (x) \supset O_1(x))]$$

in the richer language L. Then $r=3$, $b'=1$, $c'=2$, $c_o=0$ (see Fig. 4.2), whence $K-r-c_o=1>0$, $d=1$, and $2K-c'-r=3$.

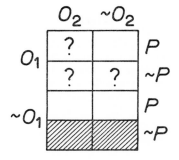

Fig. 4.2.

Now, for great α and n,

$$P(C_w/e) \simeq \frac{\left(\dfrac{\alpha}{n}\right)}{\left(1+\dfrac{\alpha}{n}\right)^2}$$

$$P(C_w/e \ \& \ T) \simeq \frac{\left(\dfrac{\alpha}{n}\right)\left(2+\dfrac{\alpha}{n}\right)}{\left(1+\dfrac{\alpha}{n}\right)^3},$$

so that $P(C_w/e) < P(C_w/e \ \& \ T) < \tfrac{1}{2}$.

For further reference, let us finally note that in Hintikka's Jerusalem system (where $\alpha=0$ and $\lambda \to \infty$), formulae (4.1)–(4.3) are replaced by

$$(4.1') \qquad P(C_w) = \frac{1}{2^K}$$

$$(4.2') \qquad P(C_w/e) = \frac{1}{\displaystyle\sum_{i=0}^{K-c}\binom{K-c}{i}\left(\dfrac{w}{c+i}\right)^n} \qquad (w = K - b)$$

$$(4.3') \qquad P(C_w/e \ \& \ T) = \frac{\displaystyle\sum_{i=0}^{d}\binom{d}{i}2^{d-i}\left(\dfrac{1}{q+i}\right)^n}{\displaystyle\sum_{i=0}^{2K-c'-r}\binom{2K-c'-r}{i}\left(\dfrac{1}{c'+i}\right)^n}.$$

NOTE

[1] The assumption $c_1 = 0$ means intuitively that the observational evidence e is as varied with respect to L and theory T as possible. This assumption is made here only for the purpose of simplifying the general formulas to be given below. However, if this assumption is not satisfied, there is no difficulty, in principle, in computing the probabilities of constituents of $L(\lambda)$ in all special cases.

CHAPTER 5

PIECEWISE DEFINABLE
THEORETICAL CONCEPTS

It has often been claimed that no gain – except possibly one in economy – results from theoretical concepts which are definable, and hence eliminable, by observational terms. Here definition normally means *explicit* definition. However, these claims do not always hold even for explicit definitions (cf. Chapter VI of Tuomela, 1973). Accordingly, they need not hold true for definitions satisfying the classical criteria of eliminability and noncreativity for definitions. An example of such a strong kind of definition satisfying these criteria is provided by a *piecewise definition* (see Hintikka and Tuomela, 1970 and Tuomela, 1973 for a discussion of its logical and methodological properties).

We say that a theoretical predicate P is *piecewise* definable in terms of the members of λ in a theory T if and only if T logically implies a finite disjunction of explicit definitions for P in terms of λ. In the special case, when this disjunction essentially contains only one disjunct, we get a case of explicit definability of P in terms of λ in T.

In this chapter, we study the inductive effects of a theory T in $L(\lambda \cup \{P\})$ that consists simply of a piecewise definition of P in terms of λ. If such a theory is found to have important inductive effects, it seems reasonable to conjecture that these effects need not in general diminish in the discussion of theories with more open theoretical concepts.

We first derive a number of results concerning piecewise definitions, and then briefly comment the methodological content of these results. It should be noted that when in Chapters 3 and 4 we studied the effect which a theory T in the richer language $L = L(\lambda \cup \{P\})$ has upon the probabilities of generalizations in $L(\lambda)$, it was assumed that theory T excludes some cells of L. This need not, however, be the case with all interesting kinds of theories or definitions. In particular, this assumption is not always satisfied in the case of piecewise definitions, that is to say, finite disjunctions of explicit definitions.

We have shown above that

(5.1)   If T is an explicit definition of P in terms of the vocabulary of $L(\lambda)$, and g is a weak or strong generalization in $L(\lambda)$, then $P(g/e) = P(g/e \,\&\, T)$ for all e in $L(\lambda)$.

Suppose now that T is equivalent to the disjunction $T_1 \lor T_2$ of two explicit definitions, T_1 and T_2, of P in terms of λ. Let t be the conjunction, $T_1 \,\&\, T_2$, of T_1 and T_2. Then t is equivalent to the disjunction of those constituents of L which occur in *both* of the distributive normal forms of T_1 and T_2. By applying result (5.1), we derive

$$(5.2) \quad P(g/e \,\&\, T) = \frac{P(g/e)\,[P(e \,\&\, T_1) + P(e \,\&\, T_2)] - P(g \,\&\, e \,\&\, t)}{P(e \,\&\, T_1) + P(e \,\&\, T_2) - P(e \,\&\, t)}.$$

If t is incompatible with the assumption that the universe concerned is non-empty (i.e., if the normal form of t contains only the constituent according to which no Ct^r-predicate is instantiated in the universe), then $P(e \,\&\, t) = P(g \,\&\, e \,\&\, t) = 0$, whence $P(g/e \,\&\, T) = P(g/e)$. Similarly, if t is incompatible with evidence e, then also $P(g/e \,\&\, T) = P(g/e)$. This happens in the event that T_1 and T_2 differ from each other in what they state about those Ct^r-predicates of L which correspond to the Ct-predicates of $L(\lambda)$ exemplified in e. (Recall that each explicit definition of P in terms of λ excludes one 'half' of all the cells of $L(\lambda)$; cf. Chapter 3.)

For strong generalizations g, t may be inconsistent with some g. This happens in the event that t rules out a Ct-predicate of $L(\lambda)$ which is instantiated by g. Formula (5.2) is then reduced to

$$(5.3) \qquad P(g/e \,\&\, T) = \frac{P(g/e)\,[P(e \,\&\, T_1) + P(e \,\&\, T_2)]}{P(e \,\&\, T_1) + P(e \,\&\, T_2) - P(e \,\&\, t)}$$

In this case,

$$P(g/e \,\&\, T) > P(g/e) \quad \text{iff} \quad P(g/e) > 0 \quad \text{and} \quad P(e \,\&\, t) > 0.$$

Consequently, we obtain the results

(5.4)   Let $T = T_1 \lor T_2$ be a piecewise definition and $t = T_1 \,\&\, T_2$. If t is incompatible with e, or if t is incompatible with the assumption that the universe is non-empty, then $P(g/e \,\&\, T) = P(g/e)$. If g is a strong generalization compatible with evidence e, and t is incompatible with g but not with e, then $P(g/e \,\&\, T) > P(g/e)$.

On the assumption that t is compatible with e, formula (5.2) implies that

(5.5) $P(g/e \& T) > P(g/e)$
iff $-P(g \& e \& t) > -P(g/e) P(e \& t)$
iff $P(g/e) > P(g/e \& t)$.

Here, probability $P(g/e)$ is obtained from formula (3.2) (if g is a weak generalization), or from formula (4.1) (if g is a strong generalization). Since t is of the same type as the theories presented in Chapters 3 and 4, probability $P(g/e \& t)$ can be computed from formula (3.3) or (4.2) (on the assumption that the predicate P is evidential).

We let

$r =$ the number of those Ct^r-predicates of L which are empty by t
$b =$ the number of those Ct-predicates of $L(\lambda)$ which are empty by g $(0 < b \leqslant K - c)$
$b' =$ the number of those Ct^r-predicates of L which are empty by g but not by t.

Now, t excludes those Ct^r-predicates which are empty according to *at least one* of the explicit definitions T_1 and T_2. Since each of these definitions excludes precisely K Ct^r-predicates, we have

$$K \leqslant r \leqslant 2K$$

(with $K = r$ just in case T_1 and T_2 are equivalent, and T thus is reduced to an explicit definition). Moreover, as t excludes at least one 'half' of each Ct-predicate of $L(\lambda)$, b' cannot exceed b. Thus,

$$0 \leqslant b' \leqslant b.$$

Suppose now that g is a weak generalization. If α and n are large, then the result (3.9) and $b' \leqslant b$ together imply that $P(g/e) \leqslant P(g/e \& t)$. Therefore, by (5.5),

(5.6) For large α and n, we always have $P(g/e \& T) \leqslant P(g/e)$, if T defines P piecewise in terms of λ and g is a weak generalization.

The general case, for any values of α and n, can be illustrated by a simple example.

Example 5.1. Let $\lambda = \{O_1, O_2\}$, $K = 4$, and $T = T_1 \vee T_2$, where

$$T_1 = (x)\,(P\,(x) \equiv O_1\,(x))$$
$$T_2 = (x)\,(P\,(x) \equiv O_2\,(x)).$$

Then $t = T_1 \,\&\, T_2$ says that the Ct^r-predicates $Ct_2^r - Ct_7^r$ are empty. (In Figure 5.1, the cells empty by T_i, $i = 1, 2$, are marked with 'T_i'.)

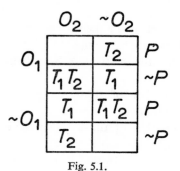

Fig. 5.1.

Let

$$g = (x)\,(\sim O_1\,(x) \supset O_2\,(x)),$$

and suppose that all individuals in the sample e belong to the Ct-predicate $Ct_1 = O_1 \,\&\, O_2$. Then $r = 6$, $c = 1$, $c' = 1$, $b = 1$, and $b' = 1$. Hence,

$$P\,(g/e) = \cfrac{1}{1 + \cfrac{\alpha + 1}{n + 1}\,\cdot\,\cfrac{1 + 2\,\dfrac{\alpha + 2}{n + 2} + \dfrac{(\alpha + 2)\,(\alpha + 3)}{(n + 2)\,(n + 3)}}{1 + 2\,\dfrac{\alpha + 1}{n + 1} + \dfrac{(\alpha + 1)\,(\alpha + 2)}{(n + 1)\,(n + 2)}}}$$

and

$$P\,(g/e \,\&\, t) = \cfrac{1}{1 + \dfrac{\alpha + 1}{n + 1}}.$$

These computations provide a result which is similar to the example of Chapter 3, viz. now $P\,(g/e \,\&\, T) > P\,(g/e)$ iff $P\,(g/e) > P\,(g/e \,\&\, t)$ iff $n < \alpha$ iff $P\,(g/e) < \frac{1}{2}$.

Suppose then that g is a strong generalization C_w. Let us first assume that $w=K-b=c$. Then by formula (4.1),

$$P(C_w/e) = \frac{\dfrac{(\alpha+c-1)!}{(n+c-1)!}}{\displaystyle\sum_{i=0}^{K-c} \binom{K-c}{i} \dfrac{(\alpha+c+i-1)!}{(n+c+i-1)!}}.$$

If T_1 and T_2 are not equivalent, and if t is compatible with e (so that t is compatible with C_w), then $b-b'>0$, $c_o=0$, $c=c'$, $d=0$, and $r= =K+b-b'>K$. Moreover, $q=K-b+c_o=c$. Hence, by formula (4.3),

$$P(C_w/e \ \& \ t) = \frac{\dfrac{(\alpha+c-1)!}{(n+c-1)!}}{\displaystyle\sum_{i=0}^{b'} \binom{b'}{i} \dfrac{(\alpha+c+i-1)!}{(n+c+i-1)!}},$$

which is always greater than $P(C_w/e)$ since $K-c=b>b'$. Consequently, in case $K-b=c$, we always have

$$P(C_w/e \ \& \ T) < P(C_w/e),$$

for piecewise definitions such that t is compatible with e.

Suppose next that $w=K-b>c$. If T_1 and T_2 are not equivalent, then $K<r\leqslant 2K$. If t is compatible with e, then $c_o=0$, i.e., $c'=c$. If now $b=b'$, then (because $r>K$ and $c'=c$) t has to rule out some Ct-predicates of $L(\lambda)$ which are not exemplified in e but are claimed to be instantiated by C_w. In other words, $b=b'$ implies that t is incompatible with C_w. Hence, $P(C_w \ \& \ e \ \& \ t)=0$. Now, by the result (5.4),

$$P(C_w/e \ \& \ T) > P(C_w/e) \quad \text{iff} \quad P(e \ \& \ t)>0.$$

However, $P(e \ \& \ t)>0$ is true on the assumption that t is compatible with e. We thus obtain the result

(5.7)   If $T=T_1 \vee T_2$ is a piecewise definition so that T_1 and T_2 are not equivalent, and $T_1 \ \& \ T_2$ is compatible with the evidence e, then $b=b'$ implies that $P(C_w/e \ \& \ T) > P(C_w/e)$ in case $w>c$. In case $w=c$, we always have $P(C_w/e \ \& \ T) < P(C_w/e)$.

This general result can be illustrated by the following simple example.

Example 5.2. Let $\lambda = \{O_1, O_2\}$, $K=4$, and $T=T_1 \vee T_2$, where

$$T_1 = (x) \, (P \, (x) \equiv O_1 (x))$$
$$T_2 = (x) \, (P \, (x) \equiv O_1 (x) \, \& \, O_2 (x)).$$

(SeeFigure 5.2.)

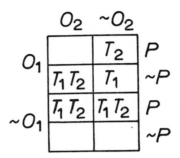

Fig. 5.2.

Let e state that $Ct_1 = O_1 \, \& \, O_2$ is exemplified, and let C_w state that $Ct_4 = = \sim O_1 \, \& \, \sim O_2$ (and only Ct_4) is empty. Then $c=1$ and $w=3$. Now t implies that $Ct_2 = O_1 \, \& \, \sim O_2$ is empty. Hence, t is incompatible with C_w. However, t is not incompatible with e and, moreover, $b=b'=1$. Consequently, by virtue of the result (5.7), we have $P \, (C_w/e \, \& \, T) > P \, (C_w/e)$.

The results established above serve to illustrate the relation between explicit and piecewise definitions. Our result (5.1) shows that the introduction of new predicates which are explicitly definable in terms of the predicates of the old language $L(\lambda)$ cannot change the inductive probabilities of generalizations in $L(\lambda)$. As can be seen from result (5.4), certain kinds of piecewise definitions share this property with explicit definitions. However, Example 5.1 and result (5.7) show that certain other kinds of piecewise definitions can increase the probabilities of generalizations in $L(\lambda)$. Consequently, in contrast with explicit definitions, piecewise definitions can have inductive effects on generalizations in $L(\lambda)$ – at least whenever these effects are characterizable in terms of an increase in probabilities (cf. Chapter 6, Chapter 7.4, and Chapter 9).

We have discussed here the definability of theoretical concepts in terms of the observational ones. It should be noted that our formal results apply also to the cases where the *observational* concepts can be thought of as being definable in terms of the *theoretical* ones (cf. Ramsey, 1931; Braithwaite, 1953; Sellars, 1963; Carnap, 1966; Hintikka, 1973, and Tuomela, 1972). Thus, suppose that T is a fixed theory in the theoretical language $L(\mu)$, and that this 'pure' theory is to be applied to account for or to explain observable phenomena characterizable in terms of the predicates λ. To do this, one has to interprete (or reinterprete) the observational predicates in λ by suitable 'correspondence rules' which serve to subordinate the observational generalizations in $L(\lambda)$ under theory T. Hintikka (1973) argues that, within such situations, one should seek for correspondence rules which do not amplify theory T in the sense of having consequences concerning the theoretical language $L(\mu)$ over and above what theory T already implies concerning $L(\mu)$. Such amplification of a theory is often said to be *ad hoc*. Corresponding to this idea, one can define a degree of *adhockery* which, as Hintikka (1973) points out, is zero for correspondence rules consisting of explicit definitions of the observational concepts λ in terms of the theoretical concepts μ.

Our results provide a kind of *inductive* analogue of Hintikka's notion of the degree of adhockery. (Hintikka's notion is concerned with deductive systematization with respect to μ.) Suppose that T is a 'pure' theory in $L(\mu)$, and that d is a conjunction of correspondence rules stated in $L(\lambda \cup \mu)$. Suppose further that P_0 and P_1 are probability measures defined for languages $L(\mu)$ and $L(\lambda \cup \mu)$, respectively. Let h be any statement of $L(\mu)$. Then the expression

$$(5.8) \qquad |P_1(h/T \ \& \ d) - P_0(h/T)|$$

can be said to measure the degree of inductive adhockery of $T \ \& \ d$ with respect to h, given T. Correspondence rules d which make (5.8) positive add something to the inductive content of T with respect to $L(\mu)$ (and are thus *ad hoc* in the situations where T is conceived of as a fixed theory which should not be amplified when it is applied to the reinterpretation and explanation of observational statements). When our formal results are applied to this situation, it follows that, within our framework, both explicit definitions and certain kinds of piecewise definitions of λ in terms of μ make the difference (5.8) zero.

CHAPTER 6

EPISTEMIC UTILITIES
AND INDUCTIVE SYSTEMATIZATION

A number of philosophers, notably Carl G. Hempel, Isaac Levi, Jaakko Hintikka, Risto Hilpinen, and Juhani Pietarinen, have recently suggested that many important aspects of scientific inference can be conceptualized in decision-theoretic terms.[1] In particular, it has been suggested that the acceptance and rejection of scientific hypotheses can be viewed as a process of maximatization of certain *epistemic utilities* that are related to the cognitive objectives of the scientist *qua* scientist. Such epistemic utilities include *truth, information, systematic* or *explanatory power*, and *simplicity*.

The theory of semantic information, which has led to many valuable insights in the studies of the philosophers mentioned above, will be applied in this chapter as a general framework under which can be discussed a variety of situations belonging to the wide area of inductive systematization. The theoreticians referred to above have restricted their discussion to observational languages or to one fixed language. However, we here extend measures of information and systematic power to an enriched language L containing theoretical predicates. As will be seen in this and later chapters, these extended measures are also intimately connected with the notions of positive relevance and degree of corroboration. There will also be established, in this chapter and in Chapter 7.4, a few results based on these measures and on our earlier technical results obtained in respect of Hintikka's inductive logic.

1. MEASURES OF INFORMATION AND SYSTEMATIC POWER

1.1. There are two different measures of the semantic information of a sentence h belonging to a fixed scientific language:

(6.1) $inf(h) = - log_2 P(h)$

and

(6.2) $cont(h) = 1 - P(h)$.

Measure *inf* is also commonly used in statistical communication theory (cf. Shannon and Weaver, 1949), while the measure *cont* was first introduced in Carnap and Bar-Hillel (1952). It has been suggested that (6.1) is a measure of the *surprise value* or the *unexpectedness* of the truth of *h*, while (6.2) is a measure of the *substantive information* or the *content* of *h* (cf. Bar-Hillel, 1964, p. 307). The fundamental difference between these two measures is reflected in their additivity conditions, viz.,

$$cont\,(h_1\,\&\,h_2) = cont\,(h_1) + cont\,(h_2)$$

if and only if $h_1 \vee h_2$ is logically true, while

$$inf\,(h_1\,\&\,h_2) = inf\,(h_1) + inf\,(h_2)$$

if and only if h_1 and h_2 are probabilistically independent.

These measures can be generalized in various respects. Thus, one may define measures of conditional, incremental, and transmitted information. We here follow the definitions of Hintikka (1968b) in which work the philosophical interpretation of these measures is discussed in more detail.

Measures of *conditional* information, i.e., the information of *h* given a statement *e*, are given by

$$(6.3) \qquad inf\,(h/e) = -\log_2 P(h/e)$$

and

$$(6.4) \qquad cont\,(h/e) = 1 - P(h/e).$$

The measures of *incremental* information can be defined as follows:

$$(6.5) \qquad inf\,(hIe) = inf\,(h\,\&\,e) - inf\,(e)$$

$$(6.6) \qquad cont\,(hIe) = cont\,(h\,\&\,e) - cont\,(e).$$

Formulae (6.5) and (6.6) express how much information *h* adds to the information that *e* has already supplied. It proves that

$$inf\,(hIe) = -\log_2 P(h/e) = inf\,(h/e)$$
$$cont\,(hIe) = cont\,(e \supset h),$$

so that, in general, $cont\,(hIe)$ and $cont\,(h/e)$ are different.

As measures of *transmitted* information we can define

(6.7) $inf(h//e) = inf(e) - inf(eIh) = inf(e) - inf(e/h)$

(6.8) $cont(h//e) = cont(e) - cont(eIh)$

(6.9) $cont(h///e) = cont(e) - cont(e/h).$

Formulae (6.7) and (6.8) express the amount of information h carries concerning the subject matter of e, i.e., the reduction of uncertainty concerning e which results when we come to know h. (6.9) measures the change in the content of e which takes place when we come to know h. Formulae (6.7)–(6.9) are reducible to

$$inf(h//e) = log_2(P(e/h)/P(e))$$
$$cont(h//e) = cont(h \vee e) = 1 - P(h \vee e)$$
$$cont(h///e) = P(e/h) - P(e).$$

Consequently, $inf(h//e)$ and $cont(h//e)$ are symmetric with respect to h and e. These measures of transmitted information can be normalized in the following way:

(6.10) $inf_e(h//e) = \dfrac{inf(h//e)}{inf(e)} = \dfrac{log\,P(e) - log\,P(e/h)}{log\,P(e)}$

(6.11) $cont_e(h//e) = \dfrac{cont(h//e)}{cont(e)} = \dfrac{1 - P(h \vee e)}{1 - P(e)} = P(\sim h/\sim e)$

(6.12) $cont_e(h///e) = \dfrac{cont(h///e)}{cont(e)} = \dfrac{P(e/h) - P(e)}{1 - P(e)}.$

Let $\mathbf{H} = \{h_1, ..., h_n\}$ be a finite set of n mutually exclusive and jointly exhaustive hypotheses h_i. We can then define *entropy*-measures analogously to formulae (6.1)–(6.9):

(6.13) $inf(\mathbf{H}) = \sum_i P(h_i)\,inf(h_i) = -\sum_i P(h_i)\,log\,P(h_i)$

(6.14) $cont(\mathbf{H}) = \sum_i P(h_i)\,cont(h_i) = 1 - \sum_i P(h_i)^2$

(6.15) $inf(\mathbf{H}/e) = inf(\mathbf{H}Ie) = -\sum_i P(h_i/e)\,log\,P(h_i/e)$

(6.16) $cont(\mathbf{H}/e) = \sum_i P(h_i/e)\,cont(h_i/e) = 1 - \sum_i P(h_i/e)^2$

(6.17) $cont(\mathbf{H}Ie) = \sum_i P(h_i/e)\,cont(h_iIe) = P(e)\,cont(\mathbf{H}/e)$

(6.18) $inf(e//\mathbf{H}) = inf(\mathbf{H}) - inf(\mathbf{H}Ie)$

(6.19) $cont(e//\mathbf{H}) = cont(\mathbf{H}) - cont(\mathbf{H}|e) =$
$$= 1 - \sum_i P(h_i)^2 - P(e)\left(1 - \sum_i P(h_i/e)^2\right)$$

(6.20) $cont(e///\mathbf{H}) = cont(\mathbf{H}) - cont(\mathbf{H}/e) = \sum_i P(h_i/e)^2 - \sum_i P(h_i)^2.$

These formulae can be interpreted as measures of information concerning the set $\mathbf{H}$ (cf. e.g. Carnap and Bar-Hillel (1952) and Hilpinen (1970)). For example, $cont(e//\mathbf{H})$ is a measure of the substantive information e carries about the set $\mathbf{H}$. Again, the measures (6.18)–(6.20) of transmitted information can be normalized:

(6.21) $inf_{\mathbf{H}}(e//\mathbf{H}) = \dfrac{inf(e//\mathbf{H})}{inf(\mathbf{H})}$

(6.22) $cont_{\mathbf{H}}(e//\mathbf{H}) = \dfrac{cont(e//\mathbf{H})}{cont(\mathbf{H})}$

(6.23) $cont_{\mathbf{H}}(e///\mathbf{H}) = \dfrac{cont(e///\mathbf{H})}{cont(\mathbf{H})}.$

Definitions (6.10)–(6.12) and (6.21)–(6.23) give us three measures of the (*informational* or *systematic*) *power* of e with respect to h:

(6.24) $syst_1(e, h) = inf_h(e//h)$

(6.25) $syst_2(e, h) = cont_h(e//h)$

(6.26) $syst_3(e, h) = cont_h(e///h),$

and three measures of the power of e with respect to the set $\mathbf{H}$:

(6.27) $syst_1(e, \mathbf{H}) = inf_{\mathbf{H}}(e//\mathbf{H})$

(6.28) $syst_2(e, \mathbf{H}) = cont_{\mathbf{H}}(e//\mathbf{H})$

(6.29) $syst_3(e, \mathbf{H}) = cont_{\mathbf{H}}(e///\mathbf{H}).$

If $e \vdash h$, then $syst_i(e, h) = 1$, for $i = 1, 2, 3$. Similarly, if $e \vdash h_i$, for some h_i in $\mathbf{H}$, then $syst_i(e, \mathbf{H}) = 1$, for $i = 1, 2, 3$.

These measures can be applied in various ways, depending on the interpretation of e, h, and $\mathbf{H}$. For example, if e is an observational evidence statement and h is a hypothesis, $syst_i$, $i = 1, 2, 3$, measure the (normed) amount of information that is provided by e about h (cf. Hilpinen, 1970). But if e contains a theory T, $syst_i$, $i = 1, 2, 3$, can be thought of as measuring the *systematic* (that is, explanatory or predictive) *power* of T with respect to h or $\mathbf{H}$, in a given evidential situation. Since we intend to apply these measures for evaluation of the systematic or explanatory power of theories, we have used the symbol '*syst*' for them. Measure (6.24) has been suggested by Törnebohm (1966), and (6.25) by Hempel and Oppenheim (1948). Measures (6.24)–(6.26) have been discussed by Hintikka (1968b), Pietarinen and Tuomela (1968), and Pietarinen (1970), and measures (6.27) and (6.28) by Hilpinen (1970).[2]

1.2.   The measures which we have defined are applicable, in many ways, for definition of the *expected utility*, $E(h/e)$, of accepting h on basis of e. If we let $u(h, t, e)$ be the utility of accepting h on e when h is true, and $u(h, f, e)$ that utility when h is false, then $E(h/e)$ is defined by

$$(6.30) \qquad E(h/e) = P(h/e)\, u(h, t, e) + P(\sim h/e)\, u(h, f, e).$$

If the utilities are measured by information, then the simplest (and the most natural) suggestion (cf. Hintikka and Pietarinen, 1966) is to define

$$(6.31) \qquad \begin{aligned} u(h, t, e) &= cont(h) \\ u(h, f, e) &= - cont(\sim h). \end{aligned}$$

(6.30) and (6.31) yield

$$(6.32) \qquad E(h/e) = P(h/e) - P(h).$$

(6.32) results also if we replace (6.31) by

$$(6.33) \qquad \begin{aligned} u(h, t, e) &= cont(h//e) \\ u(h, f, e) &= - cont(\sim h//e). \end{aligned}$$

However, $E(h/e)$ is identically zero if we measure the utilities, in a similar way, by means of incremental or conditional content.

Let $u(h, t, e, T)$ be the utility of accepting h on the basis of e and a theory T when h is true, and $u(h, f, e, T)$ that utility when h is false. Then the expected utility, $E(h/e \,\&\, T)$, of accepting h on the basis of e and T, is given by

$$(6.34) \qquad \begin{aligned} E(h/e \,\&\, T) = P(h/e \,\&\, T)\, u(h, t, e, T) + \\ + P(\sim h/e \,\&\, T)\, u(h, f, e, T). \end{aligned}$$

If we now define

$$(6.35) \quad \begin{aligned} u(h, t, e, T) &= cont\,(h) \\ u(h, f, e, T) &= -\,cont\,(\sim h), \end{aligned}$$

then

$$(6.36) \quad E(h/e \ \& \ T) = P(h/e \ \& \ T) - P(h).$$

(6.36) also results if we replace (6.35) by

$$(6.37) \quad \begin{aligned} u(h, t, e, T) &= cont\,(h//(e \ \& \ T)) \\ u(h, f, e, T) &= -\,cont\,(\sim h//(e \ \& \ T)). \end{aligned}$$

However, if the utilities are defined by

$$(6.38) \quad \begin{aligned} u(h, t, e, T) &= cont\,(h/T) \\ u(h, f, e, T) &= -\,cont\,(\sim h/T), \end{aligned}$$

then we have, instead of (6.36),

$$(6.39) \quad E(h/e \ \& \ T) = P(h/e \ \& \ T) - P(h/T).$$

(6.39) also results if we measure utilities by transmitted information $cont\,(h//e)$ relative to T, i.e.,

$$(6.40) \quad \begin{aligned} u(h, t, e, T) &= cont\,(h \vee e/T) \\ u(h, f, e, T) &= -\,cont\,(\sim h \vee e/T). \end{aligned}$$

By formulae (6.32), (6.36), and (6.39), we have three definitions for the expected utility of accepting h on the basis of e or on the basis of e and T:

$$(6.41) \quad U(h/e) = P(h/e) - P(h)$$

$$(6.42) \quad U_1\,(h/e \ \& \ T) = P(h/e \ \& \ T) - P(h)$$

$$(6.43) \quad U_2\,(h/e \ \& \ T) = P(h/e \ \& \ T) - P(h/T).$$

Thus we see that the measure U can be extended, in two different ways, to cases in which we rely on a background theory T.

These definitions have a natural interpretation in terms of positive relevance: $U(h/e) > 0$ if and only if e is positively relevant to h; $U_1\,(h/e \ \& \ T) > 0$ if and only if '$e \ \& \ T$' is positively relevant to h; and $U_2\,(h/e \ \& \ T) > 0$ if and only if e is positively relevant to h relative to T.

2. Expected Epistemic Utilities of Generalizations

2.1.   Suppose now that g is a generalization in the observational language $L(\lambda)$ and that e is the description of our evidence in $L(\lambda)$. Let T be a theory in the language $L = L(\lambda \cup \{P\})$. Then we can define, by formulae (6.41)–(6.43),

(6.44) $U(g/e) = P(g/e) - P(g)$

(6.45) $U_1(g/e \ \& \ T) = P(g/e \ \& \ T) - P(g)$

(6.46) $U_2(g/e \ \& \ T) = P(g/e \ \& \ T) - P(g/T),$

in which all the relevant probabilities can be computed by the formulae given in earlier chapters. Here $U(g/e)$ is the expected utility of accepting g on the basis of e in the observational language $L(\lambda)$. As is observable from the utility assignment (6.35), in cases where our objective is that of gaining the content of g as measured in $L(\lambda)$, the expected utility of accepting g on the basis of e and T is given by $U_1(g/e \ \& \ T)$ (where $P(g)$ is the *a priori* probability of g in $L(\lambda)$). This is the case, for example, when the theory T is regarded only as an instrument for the establishment of observational conclusions. However, as can be seen from the utility assignment (6.38), in cases where we view the acceptance of g in the light of the theory T so that our objective is that of gaining the content of g relative to T, the expected utility of accepting g on the basis of e and T is given by $U_2(g/e \ \& \ T)$ (where $P(g/T)$ is the probability of g in L relative to T). Thus, (6.45) seems to be in accord with the methodological instrumentalist's view of scientific theories, while (6.46) is often more natural for a scientific realist.

Formulae (6.44)–(6.46) can be relativized to an evidential situation or background knowledge, as expressed by a sentence b in $L(\lambda)$, by computing all the probabilities conditional on b.

The measures U, U_1, and U_2 give rise to several interesting questions which illuminate some of the different kinds of gains which can be obtained by using theoretical terms.

2.2.   In the first place, we can compare the values of $U(g/e)$, $U_1(g/e \ \& \ T)$, and $U_2(g/e \ \& \ T)$. In particular, we can ask when the expected utility of accepting an observational generalization g on the basis of evidence e and a theory T is greater than on the basis of e alone, that is to say, when we have

$$U(g/e) < U_1(g/e \ \& \ T)$$

and

$$U(g/e) < U_2(g/e \ \& \ T).$$

When these inequalities hold, the differences $U_1(g/e \ \& \ T) - U(g/e)$ and $U_2(g/e \ \& \ T) - U(g/e)$ can be regarded as tentative measures of the gains in expected utilities arising from the introduction of the new predicate 'P' and the theory T.

By (6.44)–(6.46), we see that

(6.47) $U(g/e) < U_1(g/e \ \& \ T)$ iff $P(g/e) < P(g/e \ \& \ T)$

(6.48) $U(g/e) < U_2(g/e \ \& \ T)$ iff
$$P(g/e) - P(g) < P(g/e \ \& \ T) - P(g/T).$$

Here $P(g/e)$ can be computed by (3.2) or (4.2), $P(g/e \ \& \ T)$ by (3.3), (3.17) or (4.3), $P(g)$ by (3.4) or (4.1), and $P(g/T)$ by (3.5) or by (4.3) putting $n=0$ and $c'=0$.

In Chapters 3, 4, and 5 we have established a number of results concerned with the relation between the probabilities $P(g/e)$ and $P(g/e \ \& \ T)$. In view of (6.47), all of these results can be reformulated in terms of U and U_1. To summarize the most important ones, suppose first that T is a theory which rules out some cells of the language $L = L(\lambda \cup \{P\})$ and 'P' is evidential. Then

(6.49)   Let g be a weak or strong generalization in $L(\lambda)$. Then any of the following conditions is sufficient for
$$U(g/e) = U_1(g/e \ \& \ T)$$
to hold:
(i) $n \to \infty$ and $\alpha \neq \infty$
(ii) $\alpha \to \infty$
(iii) T is an explicit definition of 'P' in terms of λ
(iv) $\alpha = n$ and $b = b'$
(v) α and n are great, and $b = b'$.

(6.50)   If α and n are great, or $\alpha = n$, then
$$U(g/e) < U_1(g/e \ \& \ T)$$
holds for a weak generalization g iff $b > b'$ (and only if T has observational consequences).

(6.51)   If α and n are great, then
$$U(g/e) < U_1(g/e \,\&\, T)$$
holds for a strong generalization g iff

$$\left(1 + \frac{\alpha}{n}\right)^{K-r-c_0} < \left(2 + \frac{\alpha}{n}\right)^{K-r-c_0+b-b'}.$$

(6.52)   If g is a weak generalization and $b+c=K$, $b'+c'+r=2K$, and $b=b'$, then
$$U(g/e) < U_1(g/e \,\&\, T)$$
holds iff $\alpha > n$ and $c_o > 0$.

For piecewise definitions, we have the following results:

(6.53)   Let g be a weak generalization in $L(\lambda)$ and let $T=T_1 \vee T_2$ be a piecewise definition of 'P' in terms of λ. Then
(i) If the conjunction $T_1 \,\&\, T_2$ is incompatible with the evidence e, then
$$U(g/e) = U_1(g/e \,\&\, T)$$
(ii) If α and n are great, then
$$U(g/e) \geqslant U_1(g/e \,\&\, T).$$

However, as Example 5.1 shows, we can have $U(g/e) < U_1(g/e \,\&\, T)$, for weak generalizations, when $n < \alpha$.

(6.54)   Let g be a strong generalization C_w in $L(\lambda)$, and let $T=T_1 \vee T_2$ be a piecewise definition such that T_1 and T_2 are not equivalent and $T_1 \,\&\, T_2$ is compatible with e. Then
(i) If $w = c$, then $U(g/e) > U_1(g/e \,\&\, T)$
(ii) If $w > c$ and $b = b'$, then $U(g/e) < U_1(g/e \,\&\, T)$.

For measures U and U_1, these results show that there cannot be any gain in expected utilities e.g. when the new predicate is explicitly definable in terms of the old vocabulary. For weak generalizations, in fairly irregular universes (α great), we have a gain in expected utilities asymptotically (n great) only if the new predicate is observationally powerful and strongly involved with observational concepts ($b > b'$). However, in certain cases it is possible to have a gain in expected utilities, even if the theory T does not have non-tautological deductive observational consequences. Examples of this are provided by piecewise definitions and by certain preasymptotic cases ($n < \alpha$) in which the evidence shows some richness with respect to the new language ($c_0 > 0$).

By (6.45) and (6.46),

(6.55)   If $P(g) = P(g/T)$, then $U_1(g/e \ \& \ T) = U_2(g/e \ \& \ T)$.

Thus, we have the following result, corresponding to (6.49), for measures U and U_2:

(6.56)   Let g be a weak or strong generalization in $L(\lambda)$. Then any of the following conditions is sufficient for
$$U(g/e) = U_2(g/e \ \& \ T)$$
to hold:
(i) $\alpha \to \infty$
(ii) T is an explicit definition of 'P' in terms of λ
(iii) α and n are great, and $b = b'$.

In correspondence with (6.51), it can be shown that

(6.57)   If α and n are great, then
$$U(g/e) < U_2(g/e \ \& \ T)$$
holds for a strong generalization g if $b = b'$ and $K - r - c_o > 0$.

It is interesting to notice that, in contrast to (6.50), $U(g/e) < U_2(g/e \ \& \ T)$ can hold for a weak generalization g even if $b < b'$, and α and n are great. To see this, let

$$x = 1 + \frac{\alpha}{n}$$
$$y = 1 + \alpha$$
$$b' = b + d \quad (d > 0)$$

where $x < y$. Then, for great α and n,

$$U(g/e) < U_2(g/e \ \& \ T) \quad \text{iff} \quad \frac{1}{y^{b'}} - \frac{1}{y^b} < \frac{1}{x^{b'}} - \frac{1}{x^b}$$
$$\text{iff} \quad \frac{1 - x^d}{1 - y^d} < \left(\frac{x}{y}\right)^{b'}.$$

If, for example, $b'=2$, $b=1$, and $d=1$, then

$$U(g/e) < U_2(g/e \& T) \quad \text{iff} \quad \frac{1-x}{1-y} < \left(\frac{x}{y}\right)^2$$

$$\text{iff} \quad \frac{1}{n} < \left(\frac{x}{y}\right)^2$$

$$\text{iff} \quad n^2 - (\alpha^2 + 1)\, n + \alpha^2 > 0$$

$$\text{iff} \quad n < 1 \text{ or } n > \alpha^2.$$

We thus arrive at the following result:

(6.58)   If α and n are great, and $b=1$ and $b'=2$, then
$$U(g/e) < U_2(g/e \& T)$$
holds for a weak generalization g iff $n > \alpha^2$.

Examples of g, T, and e satisfying the conditions of (6.58) are easily constructed. (Take, for a simple example, $g=(x)\,(O_1(x) \supset O_2(x))$ and $T = = (x)\,(O_2(x) \supset P(x))$.)

The results (6.50), (6.51), (6.57) and (6.58) show an interesting difference in the asymptotic behaviour (n and α great) of U_1 and U_2. For strong generalizations g in $L(\lambda)$, both $U_1(g/e \& T)$ and $U_2(g/e \& T)$ can be greater than $U(g/e)$ even if T does not have observational consequences (see Example 4.1). For weak generalizations of $L(\lambda)$, $U_2(g/e \& T)$ can, but $U_1(g/e \& T)$ not, be greater than $U(g/e)$ when T does not have observational consequences. This difference is attributable to the fact that theory T alone may very well diminish the probability of g more than T diminishes it in relation to evidence e.

For piecewise definitions, the following results can be proved. Let $T=T_1 \vee T_2$ be a disjunction of two explicit definitions, and let $t=T_1 \& T_2$. Then

$$P(g/e \& T) > P(g/e) \quad \text{iff} \quad P(g/e) > P(g/e \& t)$$

and

$$P(g/T) > P(g) \quad \text{iff} \quad P(g) > P(g/t).$$

Moreover,

(6.59) $U(g/e) < U_2(g/e \& T)$ iff
$$P(e/t)\,[P(g/e) - P(g/e \& t)] > P(e/T)\,[P(g) - P(g/t)]$$

as, by (5.2),

$$P(g/e \& T) - P(g/e) =$$
$$= P(g/e)\left[\frac{P(e \& T_1) - P(e \& T_2) - P(e)P(t/e \& g)}{P(T \& e)} - 1\right]$$
$$= \frac{P(t \& e)}{P(T \& e)}[P(g/e) - P(g/e \& t)]$$

and, similarly,

$$P(g/T) - P(g) = P(g)\frac{P(t) - P(t/g)}{P(T)}$$
$$= \frac{P(t)}{P(T)}[P(g) - P(g/t)].$$

As special cases of (6.59), we have

(6.60)   If t is incompatible with e, then
$$U(g/e) < U_2(g/e \& T) \quad \text{iff} \quad P(g/t) > P(g)$$
$$\text{and} \quad P(e/T) > 0.$$

(6.61)   If t is incompatible with a strong generalization g, but not with e, then
$$U(g/e) < U_2(g/e \& T) \quad \text{iff} \quad P(e/t)P(g/e) > P(e/T)P(g).$$

(6.62)   If t is not incompatible with e, then $U(g/e) < U_2(g/e \& T)$ holds if $P(g) \leqslant P(g/t)$ and $P(g/e) > P(g/e \& t)$.

On the basis of these observations, the following results can be proved.

(6.63)   Let g be a weak generalization in $L(\lambda)$ and let $T = T_1 \vee T_2$ be a piecewise definition of 'P' in terms of λ. Then
   (i) If the conjunction $T_1 \& T_2$ is incompatible with the evidence e, then
$$U(g/e) < U_2(g/e \& T)$$
   holds in case $b' < b$ and α is sufficiently large.
   (ii) If α and n are large, and $b' = b$, then
$$U(g/e) \approx U_2(g/e \& T).$$

However, even in cases where $T_1 \& T_2$ is compatible with e, it seems that

$U(g/e) < U_2(g/e \ \& \ T)$ may be true when $b' = b$ but α and n are not too large, or when $b' < b$ but α and n are large.

(6.64)   Let g be a strong generalization C_w in $L(\lambda)$ and let $T = T_1 \vee T_2$ be a piecewise definition of 'P' in terms of λ. If $w = c$, then

(i) If $T_1 \ \& \ T_2$ is incompatible with e (and, hence, with C_w), then $U(g/e) > U_2(g/e \ \& \ T)$

(ii) If $T_1 \ \& \ T_2$ is compatible with e (and, hence, with C_w) and n is sufficiently great, then $U(g/e) < U_2(g/e \ \& \ T)$.

If $w > c$, then

(iii) If $T_1 \ \& \ T_2$ is incompatible with e (and, hence, with C_w), then $U(g/e) > U_2(g/e \ \& \ T)$

(iv) If $T_1 \ \& \ T_2$ is compatible with e but not with C_w (e.g., if $b' = b$), then $U(g/e) < U_2(g/e \ \& \ T)$ holds when α and n are great and $\alpha \simeq n$.

For the proof of (6.63) and (6.64), note first that, since here $0 \leqslant b' \leqslant b$, the conditions of (6.62) cannot hold for a weak generalization g when α and n are large. Thus, for large α and n, we have

$$P(g) \leqslant P(g/t) \quad \text{and} \quad P(g/e) \leqslant P(g/e \ \& \ t),$$

with equality if and only if $b' = b$. This implies (6.63)(ii). Moreover, in (6.60), the condition $P(e/T) > 0$ is always true, and the condition $P(g/t) > > P(g)$ holds for a weak generalization g, e.g., if $b' < b$ and α is sufficiently great. This, in turn, implies (6.63)(i).

(6.64) (i) and (iii) follow immediately from (6.60). To prove (ii), suppose that t is compatible with e and C_w. Then the difference

$$P(C_w/e) - P(C_w/e \ \& \ t)$$

approaches zero when n grows without limit. However, if T_1 and T_2 are not equivalent, then

$$P(C_w) < P(C_w/t),$$

viz.,

$$P(C_w/t) = \frac{\dfrac{(\alpha + w - 1)!}{(i - 1)!}}{\displaystyle\sum_{i=0}^{2K-r} \binom{2K - r}{i} \dfrac{(\alpha + i - 1)!}{(i - 1)!}}$$

by (4.3) with $c' = c_o = 0$, $n = 0$, $q = w$, and $d = 0$. This value is always greater than $P(C_w)$, given by (4.1), since $2K - r > K$. Moreover, when t is compatible with e,

$$P(e/t) > P(e/T)$$

holds at least when n is sufficiently great. These facts, together with (6.59), imply (6.64)(ii).

To prove (6.64)(iv), note that when α and n are great, we have by (6.62),

$$U(C_w/e) < U_2(C_w/e \ \& \ T)$$

$$\text{iff} \quad \frac{P(e/t)}{P(e/T)} > \frac{P(C_w)}{P(C_w/e)}$$

$$\text{iff} \quad \frac{2(1 + \alpha)^{r_o} - 1}{2\left(1 + \dfrac{\alpha}{n}\right)^{r_o} - 1} > \frac{\left(1 + \dfrac{\alpha}{n}\right)^{K - c}}{\left(\dfrac{\alpha}{n}\right)^{K - b - c} (1 + \alpha)^b},$$

where r_0 is the number of cells of L empty by t minus K. If $\alpha \simeq n$, so that $\alpha/n \simeq 1$, this condition is reduced to

$$(1 + \alpha)^b [2(1 + \alpha)^{r_o} - 1] > 2^{K - c} [2 \cdot 2^{r_o} - 1],$$

which holds for sufficiently large values of α.

For measures U and U_2, our results show that piecewise definitions can effect a gain in the expected utilities of both weak and strong generalizations.

2.3. Above, we have compared the values of $U_i(g/e \ \& \ T)$, $i = 1, 2$, with $U(g/e)$. Another question in regard to the behaviour of these measures is to ask when we have the situations

(6.65) $\quad U(g/e) \leqslant 0 \quad$ but $\quad U_1(g/e \ \& \ T) > 0$

and

(6.66) $\quad U(g/e) \leqslant 0 \quad$ but $\quad U_2(g/e \ \& \ T) > 0$.

In other words, when is it that the expected utility of accepting g on e is non-positive, but the expected utility of accepting g on e relative to T

(or, e and T) is positive? Now, (6.65) holds if and only if

(6.67) $P(g/e) \leqslant P(g)$ but $P(g/e \ \& \ T) > P(g)$,

and (6.66) holds if and only if

(6.68) $P(g/e) \leqslant P(g)$ but $P(g/e \ \& \ T) > P(g/T)$.

(6.67) states that e is not positively relevant to g but '$e \ \& \ T$' is, while (6.68) states that e is not positively relevant to g, but e is positively relevant to g relative to T. Consequently, according to the definitions of Chapter 1.2, asking when the conditions for (6.65) and (6.66) hold amounts to asking when T does establish *inductive systematization* between e and g *in the positive relevance sense*. In particular, cases in which (6.65) and (6.66) hold for theories T lacking observational consequences give us examples in which the theoretical term 'P' occurring in T is *logically indispensable* for inductive systematization with respect to λ.

For these reasons, the questions concerning (6.65) and (6.66) are taken up in Chapter 9, which contains a discussion of the question of logical indispensability of theoretical terms for inductive systematization. There we give simple examples in which conditions (6.65) and (6.66) are satisfied.

3. COMPETING GENERALIZATIONS

3.1.   In addition to comparison of the different values of the expected utility of one particular generalization g, we can compare the expected utilities of two *competing generalizations* g_1 and g_2. Here competing generalizations are simply defined as non-equivalent generalizations of $L(\lambda)$ both of which are compatible with the observational evidence e. Thus, we can ask when it is the case that

(6.69) $U(g_1/e) \leqslant U(g_2/e)$ but $U_1(g_1/e \ \& \ T) > U_1(g_2/e \ \& \ T)$

and when

(6.70) $U(g_1/e) \leqslant U(g_2/e)$ but $U_2(g_1/e \ \& \ T) > U_2(g_2/e \ \& \ T)$.

In cases such as these, theory T effects a change in the epistemic status, as measured by expected utilities, of the two competing generalizations g_1 and g_2. If, for example, (6.70) holds, then the expected utility of g_2

on e is greater than or equal to that of g_1, but, seen in the light of the richer language and theory T, the case is just the opposite.

According to the definitions of U, U_1, and U_2,

(6.71) $U(g_1/e) \leqslant U(g_2/e)$ iff $P(g_1/e) - P(g_1) \leqslant P(g_2/e) - P(g_2)$

(6.72) $U_1(g_1/e \& T) > U_1(g_2/e \& T)$ iff
$$P(g_1/e \& T) - P(g_1) > P(g_2/e \& T) - P(g_2)$$

(6.73) $U_2(g_1/e \& T) > U_2(g_2/e \& T)$ iff
$$P(g_1/e \& T) - P(g_1/T) > P(g_2/e \& T) - P(g_2/T).$$

We assume first that g_1 and g_2 are weak generalizations which are compatible with the evidence e, and denote by b_i and b'_i, $i = 1, 2$, the values of b and b' for g_1 and g_2, respectively. Then

(6.74) $P(g_1) < P(g_2)$ iff $P(g_1/e) < P(g_2/e)$ iff $b_1 > b_2$

(6.75) $P(g_1/T) < P(g_2/T)$ iff $P(g_1/e \& T) < P(g_2/e \& T)$
$$\text{iff} \quad b'_1 > b'_2.$$

On combination of these results, we immediately derive

(6.76)   If g_1 and g_2 are two weak generalizations in $L(\lambda)$ such that both are compatible with e and, moreover, $b_1 = b_2$ and $b'_1 < b'_2$, then

$$U(g_1/e) = U(g_2/e) \quad \text{and} \quad U_1(g_1/e \& T) > U_1(g_2/e \& T).$$

However, even if $b_1 = b_2$ implies that $U(g_1/e) = U(g_2/e)$, and $b'_1 = b'_2$ implies that $U_2(g_1/e \& T) = U_2(g_2/e \& T)$, the converse implications do not hold in general. Thus, for example, if $b'_1 < b'_2$, then $U_2(g_1/e \& T)$ may be greater than, less than, or equal to $U_2(g_2/e \& T)$. If we let

$$x = 1 + \frac{\alpha}{n}; \qquad y = 1 + \alpha; \qquad \text{and} \ b'_2 = b'_1 + d, \quad \text{where} \quad d > 0,$$

then, for great α and n,

(6.77) $U_2(g_1/e \& T) > U_2(g_2/e \& T)$ iff $\left(\dfrac{x}{y}\right)^{b'_2} < \dfrac{x^d - 1}{y^d - 1}.$

As an illustrative special case of (6.77), we have

(6.78)   If α and n are great and, moreover, $b'_1 = 1$ and $b'_2 = 2$, then
$$U_2(g_1/e \ \& \ T) > U_2(g_2/e \ \& \ T) \quad \text{iff} \quad n < \alpha^2.$$

The following examples illustrate what has been said above.

Example 6.1. Let $L(\lambda)$ have two primitive predicates, so that $k = 2$ and $K = 4$, and let

$$
\begin{aligned}
g_1 &= (x)\,(O_1(x) \ \& \ O_2(x)) \\
g_2 &= (x)\,O_1(x) \\
g_3 &= (x)\,O_2(x) \\
g_4 &= (x)\,(O_1(x) \supset O_2(x)).
\end{aligned}
$$

Then $b_1 = 3$, $b_2 = 2$, $b_3 = 2$, and $b_4 = 1$. Suppose, further, that only the Ct-predicate '$O_1x \ \& \ O_2x$' is exemplified in evidence e. Then

$$P(g_1) < P(g_2) = P(g_3) < P(g_4),$$

and

$$P(g_1/e) < P(g_2/e) = P(g_3/e) < P(g_4/e).$$

Consequently,

$$U(g_2/e) = U(g_3/e).$$

Moreover, if α and n are great, and if $n < \alpha^2$, then

$$U(g_2/e) < U(g_4/e).$$

Suppose, then, that T is a theory asserting that

$$T = (x)\,(P(x) \supset O_1(x)),$$

where 'P' is a new predicate, no matter whether evidential or non-evidential. T states, in effect, that the Ct'-predicates '$\sim O_1x \ \& \ O_2x \ \& \ Px$' and '$\sim O_1x \ \& \sim O_2x \ \& \ Px$' of the richer language are empty. Then $b'_1 = 4$, $b'_2 = 2$, $b'_3 = 3$, and $b'_4 = 2$, whence

$$P(g_1/e \ \& \ T) < P(g_3/e \ \& \ T) < P(g_2/e \ \& \ T) = P(g_4/e \ \& \ T).$$

Thus,

$$U_1(g_2/e \ \& \ T) > U_1(g_3/e \ \& \ T)$$

and

$$U_1(g_2/e \ \& \ T) > U_1(g_4/e \ \& \ T).$$

However, in this case,

$$U_2(g_2/e \,\&\, T) = U_2(g_4/e \,\&\, T),$$

while

$$U_2(g_2/e \,\&\, T) > U_2(g_3/e \,\&\, T)$$

holds only for certain values of α and n (e.g., for large α and n such that $n < \alpha^3$).

Example 6.2. Let

$$g_1 = (x)(O_1(x) \supset O_2(x))$$
$$g_2 = (x)(O_2(x) \supset O_1(x)),$$

and suppose that Ct-predicates '$O_1x \,\&\, O_2x$' and '$\sim O_1x \,\&\, \sim O_2x$' are exemplified in e. Then, $b_1 = 1$ and $b_2 = 1$. Let T be a theory to the effect that the Ct^r-predicates '$O_1x \,\&\, \sim O_2x \,\&\, \sim Px$' and '$\sim O_1x \,\&\, \sim O_2x \,\&\, \sim Px$' are empty, i.e.,

$$T = (x)(\sim O_2(x) \supset P(x)).$$

Then, $b_1' = 1$ and $b_2' = 2$. Thus,

$$U(g_1/e) = U(g_2/e),$$

and, by (6.78),

$$U_2(g_1/e \,\&\, T) > U_2(g_2/e \,\&\, T),$$

for large α and n such that $n < \alpha^2$.

3.2. We now study (6.69) and (6.70) in the event that g_1 and g_2 are strong generalizations, say C_{w_1} and C_{w_2}. In this case,

(6.79) $P(C_{w_1}) > P(C_{w_2})$ iff $w_1 > w_2$ iff $b_1 < b_2$.

(6.80) $P(C_{w_1}/e) = P(C_{w_2}/e)$ iff $w_1 = w_2$ or $\alpha = n$.

(6.81)   If $\alpha > n$, then
        $P(C_{w_1}/e) > P(C_{w_2}/e)$ iff $w_1 > w_2$ iff $b_1 < b_2$.
    If $\alpha < n$, then
        $P(C_{w_1}/e) > P(C_{w_2}/e)$ iff $w_1 < w_2$ iff $b_1 > b_2$.

(6.82)   If $b_1 = b_2$, then

$$P(C_{w_1}/T) > P(C_{w_2}/T) \quad \text{iff} \quad b'_1 < b'_2.$$
If $b_1 < b_2$, and $b_2 - b_1 \leqslant b'_2 - b'_1$, then
$$P(C_{w_1}/T) > P(C_{w_2}/T).$$

(6.83)   If $b_1 = b_2$, then
$$P(C_{w_1}/e \,\&\, T) > P(C_{w_2}/e \,\&\, T) \quad \text{iff} \quad b'_1 < b'_2.$$
In case $b_1 < b_2$,
$$P(C_{w_1}/e \,\&\, T) > P(C_{w_2}/e \,\&\, T)$$
holds if
$$\alpha = n \quad \text{and} \quad b_2 - b_1 < b'_2 - b'_1,$$
or if
$$\alpha > n \quad \text{and} \quad b_2 - b_1 \leqslant b'_2 - b'_1,$$
and
$$P(C_{w_1}/e \,\&\, T) < P(C_{w_2}/e \,\&\, T)$$
holds if
$$\alpha < n \quad \text{and} \quad b_2 - b_1 > b'_2 - b'_1.$$

These results follow immediately from formulae (4.1)–(4.3). (Note that the value of $P(C_w/T)$ is obtained from (4.3) by putting $n = 0$, $c_o = 0$, whence $q = K - b = w$ and $d = K - r + b - b'$.)

It is now observable that (6.76) also holds for strong generalizations:

(6.84)   If C_{w_1} and C_{w_2} are two constituents of $L(\lambda)$ such that both are compatible with e and, moreover, $b_1 = b_2$ and $b'_1 < b'_2$, then
$$U(C_{w_1}/e) = U(C_{w_2}/e) \quad \text{and} \quad U_1(C_{w_1}/e \,\&\, T) >$$
$$> U_1(C_{w_2}/e \,\&\, T).$$

Furthermore,

(6.85)   If $b_1 < b_2$, $b_2 - b_1 > b'_2 - b'_1$, and $\alpha \leqslant n$, then
$$U(C_{w_1}/e) < U(C_{w_2}/e) \quad \text{and}$$
$$U_1(C_{w_1}/e \,\&\, T) < U_1(C_{w_2}/e \,\&\, T).$$

However, in cases where $b_1 < b_2$ and $b_2 - b_1 \leqslant b'_2 - b'_1$, we may have $U(C_{w_1}/e) < U(C_{w_2}/e)$ and $U_i(C_{w_1}/e \,\&\, T) > U_i(C_{w_2}/e \,\&\, T)$, $i = 1, 2$.

No attempt is made here at a more general study of the conditions for (6.69) and (6.70). However, the following example suffices to illustrate a number of various aspects of the general situation.

Example 6.3. Let $L(\lambda)$ have two primitive predicates and let

$$g_1 = (Ex)\, Ct_1(x)\, \&\, (x)\, (Ct_1(x))$$
$$g_2 = (Ex)\, Ct_1(x)\, \&\, (Ex)\, Ct_2(x)\, \&\, (x)\, (Ct_1(x) \vee Ct_2(x))$$
$$g_3 = (Ex)\, Ct_1(x)\, \&\, (Ex)\, Ct_3(x)\, \&\, (x)\, (Ct_1(x) \vee Ct_3(x))$$
$$g_4 = (Ex)\, Ct_1(x)\, \&\, (Ex)\, Ct_3(x)\, \&\, (Ex)\, Ct_4(x)\, \&$$
$$\&\, (x)\, (Ct_1(x) \vee Ct_3(x) \vee Ct_4(x)),$$

where

$$Ct_1 = O_1\, \&\, O_2$$
$$Ct_2 = O_1\, \&\, {\sim} O_2$$
$$Ct_3 = {\sim} O_1\, \&\, O_2$$
$$Ct_4 = {\sim} O_1\, \&\, {\sim} O_2.$$

Suppose that only Ct_1 is exemplified in the evidence e. Now we have

$$w_1 = 1, \qquad b_1 = 3$$
$$w_2 = 2, \qquad b_2 = 2$$
$$w_3 = 2, \qquad b_3 = 2$$
$$w_4 = 3, \qquad b_4 = 1.$$

Moreover,

$$P(g_1) = \frac{\alpha!}{(\alpha - 1)! + 4\alpha! + 6(\alpha + 1)! + 2(\alpha + 2)! + \tfrac{1}{6}(\alpha + 3)!}$$
$$P(g_2) = P(g_3) = (\alpha + 1)\, P(g_1)$$
$$P(g_4) = \tfrac{1}{2}(\alpha + 1)\, (\alpha + 2)\, P(g_1),$$

and

$$P(g_1/e) = \frac{\alpha!/n!}{\dfrac{\alpha!}{n!} + 3\dfrac{(\alpha + 1)!}{(n + 1)!} + 3\dfrac{(\alpha + 2)!}{(n + 2)!} + \dfrac{(\alpha + 3)!}{(n + 3)!}}$$
$$P(g_2/e) = P(g_3/e) = \frac{\alpha + 1}{n + 1}\, P(g_1/e)$$
$$P(g_4/e) = \frac{(\alpha + 1)\, (\alpha + 2)}{(n + 1)\, (n + 2)}\, P(g_1/e).$$

Hence,

$$(6.86) \qquad P(g_1) < P(g_2) = P(g_3) < P(g_4)$$

(6.87)   If $n < \alpha$,   then
$$P(g_1/e) < P(g_2/e) = P(g_3/e) < P(g_4/e).$$
If $n = \alpha$,   then
$$P(g_1/e) = P(g_2/e) = P(g_3/e) = P(g_4/e) = \tfrac{1}{8}.$$
If $n > \alpha$,   then
$$P(g_1/e) > P(g_2/e) = P(g_3/e) > P(g_4/e).$$

Suppose, further, that T is a theory

$$T = (x)(P(x) \supset O_1(x))$$

in the language L which contains 'O_1', 'O_2', and 'P' as primitive predicates, and that the individuals in e exemplify both 'O_1 & O_2 & P' and 'O_1 & O_2 & $\sim P$' (see Figure 6.1).

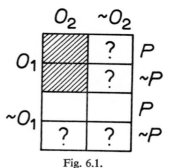

Fig. 6.1.

Now the probabilities $P(g_i/e \ \& \ T)$, $i=1, 2, 3, 4$, can be computed from (4.3) by substituting the values $K=4$, $r=2$, $c_o=1$, $c'=2$, and

$$
\begin{array}{lll}
b_1' = 4, & q_1 = 2, & d_1 = 0 \\
b_2' = 2, & q_2 = 3, & d_2 = 1 \\
b_3' = 3, & q_3 = 3, & d_3 = 0 \\
b_4' = 2, & q_4 = 4, & d_4 = 0.
\end{array}
$$

Similarly, the probabilities $P(g_i/T)$, $i=1, 2, 3, 4$, can be computed from (4.3) by substituting the values $K=4$, $r=2$, $c_o=0$, $c'=0$, $n=0$, b_i' as above, q_i one less than above, and d_i one greater than above ($i=1, 2, 3, 4$). If we let

$$\frac{1}{A} = \sum_{i=0}^{6} \binom{6}{i} \frac{(\alpha + i - 1)!}{(i-1)!}$$

and

$$\frac{1}{B} = \sum_{i=0}^{4} \binom{4}{i} \frac{(\alpha + i + 1)!}{(n + i + 1)!},$$

then

$$P(g_1/T) = A[2\alpha! + (\alpha + 1)!]$$
$$P(g_2/T) = A[4(\alpha + 1)! + 2(\alpha + 2)! + \tfrac{1}{6}(\alpha + 3)!]$$
$$P(g_3/T) = A[2(\alpha + 1)! + \tfrac{1}{2}(\alpha + 2)!]$$
$$P(g_4/T) = A[(\alpha + 2)! + \tfrac{1}{6}(\alpha + 3)!]$$

and

$$P(g_1/e \ \& \ T) = B \frac{(\alpha + 1)!}{(n + 1)!}$$

$$P(g_2/e \ \& \ T) = \left[2\frac{\alpha + 2}{n + 2} + \frac{(\alpha + 2)(\alpha + 3)}{(n + 2)(n + 3)} \right] P(g_1/e \ \& \ T)$$

$$P(g_3/e \ \& \ T) = \frac{(\alpha + 2)}{(n + 2)} P(g_1/e \ \& \ T)$$

$$P(g_4/e \ \& \ T) = \frac{(\alpha + 2)(\alpha + 3)}{(n + 2)(n + 3)} P(g_1/e \ \& \ T).$$

Hence,

(6.88) $P(g_1/T) < P(g_3/T) < P(g_4/T) < P(g_2/T).$

(6.89)   If $n < \alpha$, then
$$P(g_1/e \ \& \ T) < P(g_3/e \ \& \ T) < P(g_4/e \ \& \ T) <$$
$$< P(g_2/e \ \& \ T).$$
If $n = \alpha$, then
$$P(g_1/e \ \& \ T) = P(g_3/e \ \& \ T) = P(g_4/e \ \& \ T) <$$
$$< P(g_2/e \ \& \ T).$$
If $n > \alpha$, then
$$P(g_i/e \ \& \ T) > P(g_3/e \ \& \ T) > P(g_4/e \ \& \ T), \text{ for } i=1, 2,$$
and
$$P(g_1/e \ \& \ T) > P(g_2/e \ \& \ T)$$
iff
$$n > \tfrac{1}{2}[2\alpha - 1 + \sqrt{8\alpha^2 + 40\alpha + 49}].$$

By virtue of these results, we can now compare the expected utilities of g_1, g_2, g_3, and g_4. Suppose first that $n \geqslant \alpha$. Then

$$U(g_1/e) > U(g_2/e) = U(g_3/e) > U(g_4/e).$$

However, for U_1 we have

$$U_1(g_i/e \& T) > U_1(g_3/e \& T) >$$
$$> U_1(g_4/e \& T), \quad \text{for} \quad i = 1, 2,$$

and, for suitable values of n and α,

$$U_1(g_2/e \& T) > U_1(g_1/e \& T).$$

Thus, we see that while g_1 has the greatest expected utility, when measured by U, g_2 may have the greatest expected utility, when measured by U_1. A similar conclusion also holds for the measure U_2.

Suppose then that $n < \alpha$. Then, depending on the values of n and α, we may have

$$U(g_4/e) > U(g_1/e) > U(g_2/e) = U(g_3/e)$$

or

$$U(g_4/e) > U(g_2/e) = U(g_3/e) > U(g_1/e).$$

However, for U_1 we have

$$U_1(g_2/e \& T) > U_1(g_i/e \& T), \quad \text{for} \quad i = 3, 4.$$

These simple comparisons are adequate to indicate how theory T favours g_2 *vis-à-vis* g_1, g_3, and g_4. The generalizations g_2 and g_3 are 'symmetric' with respect to the evidence e in $L(\lambda)$, which is reflected in the equality $U(g_2/e) = U(g_3/e)$, for all n and α. However, when n is not too great when compared to α, g_2 may have the greatest expected utility among the generalizations g_i, $i = 1, 2, 3, 4$, on the basis of e and T. But, of course, if n grows without limit and α is finite, only g_1 will have positive expected utility on e, or on e and T.

This example indicates how a theory employing theoretical concepts can be used to create *asymmetries* between generalizations which are symmetric with respect to observational evidence. This important feature of our framework is also discussed in Chapters 9–11.

NOTES

[1] See Hempel (1960) and (1962), Levi (1967a, b), Hintikka and Pietarinen (1966), Hintikka (1968b), Hilpinen (1968), (1970), and (1972), and the further references given in these works.

[2] For other discussions of explanatory power and of notions related to it, see Popper (1959), Good (1960). Hanna (1966), Greeno (1970), Rosenkrantz (1970), Kyburg (1970), and Jeffrey (1971).

CHAPTER 7

THEORETICAL CONCEPTS
AND INDUCTIVE EXPLANATION

1. EXPLANATORY POWER OF THEORIES

In this chapter, we discuss the inductive explanation of (non-probabilistic) generalizations (or laws) by means of (non-probabilistic) theories. Our general viewpoint is that inductive explanation, and indeed a major part of inductive systematization, is best conceived as information-providing argumentation, but it is argumentation that does not lead to detaching a conclusion from some premises. Thus, for instance, to explain something inductively is to give a certain amount of information relevant to the explanandum. In other words, an explanans is assumed to convey some information concerning the explanandum. The more the explanans carries such relevant information, the better the resulting explanation is. As in this chapter information will generally be measured in terms of probabilities (for instance, by positive relevance), we can alternatively say that within inductive systematization, the explanation of something is stating some propositions inductively relevant to the explanandum.[1]

We have earlier defined some measures of *systematic* or *explanatory power* based upon transmitted information (see Chapter 6). Before these are discussed in more detail, a methodological remark is in order. One can distinguish on two grounds between *local* and *global* measures of explanatory power. A measure may be local or global both with respect to the explanandum and with respect to the explanans. A measure which is local with respect to the explanandum measures the explanatory power of an explanans with respect to some particular member (e.g. singular or general statement) of a totality (e.g. of the set of all observational statements) whereas a global measure measures the overall explanatory power of the explanans with respect to that totality. Both local and global measures (in this sense) are discussed below.

A measure of explanatory power is – speaking rather loosely – said to be local with respect to the explanans if the explanans is meant to (information theoretically) 'cover' the explanandum as well as possible, but not to go (much) beyond the explanandum. On the other hand, it is global with respect to the explanans if the emphasis is on the potential and intrinsic explanatory strength of the explanans rather than on its explanatory power with respect to some particular given explanandum (be it local or global). Our measures below also include both local and global ones in this sense.[2]

With this information-theoretical starting point, we proceed to discussion of the measurement of the explanatory power of theories (stated in the language $L(\lambda \cup \mu)$ introducing a set μ of theoretical concepts) with respect to observational generalizations in $L(\lambda)$. Let our explanandum be called g. Usually, it will be assumed that g is a weak or strong generalization in $L(\lambda)$ compatible with e. Then the systematic power of a theory T in $L(\lambda \cup \{P\})$, with some additional evidence e in $L(\lambda)$, with respect to g can be defined, in accordance with formulae (6.24)–(6.26), by

$$(7.1) \qquad syst_1 (T \,\&\, e, g) = \frac{\log P (g) - \log P (g/e \,\&\, T)}{\log P (g)}$$

$$(7.2) \qquad syst_2 (T \,\&\, e, g) = \frac{1 - P ((e \,\&\, T) \vee g)}{1 - P (g)} = P (\sim (e \,\&\, T)/\sim g)$$

$$(7.3) \qquad syst_3 (T \,\&\, e, g) = \frac{P (g/e \,\&\, T) - P (g)}{1 - P (g)}.$$

Instead of (7.1)–(7.3), we could define the systematic power of T with respect to g *relative to* e by

$$(7.4) \qquad syst'_1 (T \,\&\, e, g) = \frac{\log P (g/e) - \log P (g/e \,\&\, T)}{\log P (g/e)}$$

$$(7.5) \qquad syst'_2 (T \,\&\, e, g) = \frac{1 - P (T \vee g/e)}{1 - P (g/e)} = P (\sim T/\sim g \,\&\, e)$$

$$(7.6) \qquad syst'_3 (T \,\&\, e, g) = \frac{P (g/e \,\&\, T) - P (g/e)}{1 - P (g/e)}.$$

All these measures have the value one if $(T \,\&\, e) \vdash g$. Moreover,

$$syst_1 (T \,\&\, e, g) > 0 \quad \text{iff} \quad P (g) < P (g/e \,\&\, T)$$
$$syst_2 (T \,\&\, e, g) > 0 \quad \text{iff} \quad P ((e \,\&\, T) \vee g) < 1$$
$$syst_3 (T \,\&\, e, g) > 0 \quad \text{iff} \quad P (g) < P (g/e \,\&\, T)$$

(7.7) $syst'_1(T \,\&\, e, g) > 0$ iff $P(g/e) < P(g/e \,\&\, T)$
$syst'_2(T \,\&\, e, g) > 0$ iff $P(T \lor g/e) < 1$
$syst'_3(T \,\&\, e, g) > 0$ iff $P(g/e) < P(g/e \,\&\, T)$.

In other words, $syst_1(T \,\&\, e, g)$ and $syst_3(T \,\&\, e, g)$ are positive if and only if '$e \,\&\, T$' is positively relevant to g; $syst'_1(T \,\&\, e, g)$ and $syst'_3(T \,\&\, e, g)$ are positive if and only if T is positively relevant to g relative e. Several results in relation to these conditions have been established earlier.

An interesting problem in regard to the measures (7.1)–(7.6) arises on comparison of the systematic power of different theories T_1 and T_2 with respect to g. In general, we have

$syst_1(T_1 \,\&\, e, g) < syst_1(T_2 \,\&\, e, g)$
iff $syst_3(T_1 \,\&\, e, g) < syst_3(T_2 \,\&\, e, g)$
(7.8)   iff $syst'_1(T_1 \,\&\, e, g) < syst'_1(T_2 \,\&\, e, g)$
iff $syst'_3(T_1 \,\&\, e, g) < syst'_3(T_2 \,\&\, e, g)$
iff $P(g/e \,\&\, T_1) < P(g/e \,\&\, T_2)$.

The relation

(7.9) $P(g/e \,\&\, T_1) < P(g/e \,\&\, T_2),$

which can be called the *likelihood condition*, also to be called the *supersessance relation* (with respect to systematization or explanation), is studied in Section 2 below. Furthermore,

(7.10) $syst_2(T_1 \,\&\, e, g) < syst_2(T_2 \,\&\, e, g)$ iff
$syst'_2(T_1 \,\&\, e, g) < syst'_2(T_2 \,\&\, e, g)$ iff
$P(e \,\&\, T_1)\,[1 - P(g/e \,\&\, T_1)] > P(e \,\&\, T_2)\,[1 - P(g/e \,\&\, T_2)]$ iff
$P(T_1/e)\,[1 - P(g/e \,\&\, T_1)] > P(T_2/e)\,[1 - P(g/e \,\&\, T_2)].$

If $P(T_1/e) = P(T_2/e)$, (7.10) gives the likelihood relation (7.9). Note that in a sense $P(T/e)$ measures the *lack of boldness* of a theory T. That is, if e expresses our background information (be it purely empirical or not), then $cont(T/e) = 1 - P(T/e)$ measures the *boldness* of T or the excess information contained in T (relative to e). In the special case $P(g/e \,\&\, T_1) = P(g/e \,\&\, T_2) \neq 1$, (7.9) reduces to $P(T_1/e) > P(T_2/e)$, i.e. to $cont(T_1/e) > cont(T_2/e)$, and boldness becomes the only differentiating factor. The measures $syst_2$ and $syst'_2$ seem methodologically interesting just because they account for both the likelihood aspect (local explanans-factor) and the boldness aspect (global explanans-property).

Our results (7.8) and (7.10) show that no important difference exists between the 'absolute' measures $syst_i$ and their relativized counterparts $syst_i'$.

Suppose next that $\mathbf{D}$ is the set of all constituents of the observational language $L(\lambda)$. We then ask how much a theory T reduces our ignorance or uncertainty concerning this totality $\mathbf{D}$, given the evidence e. The systematic power of a theory T in $L(\lambda \cup \{P\})$, with some additional evidence e, with respect to $\mathbf{D}$, i.e., with respect to observational generalizations, can be measured, according to formulae (6.27)–(6.29), by

$$(7.11) \quad syst_1 (T \,\&\, e, \mathbf{D}) = \frac{inf(\mathbf{D}) - inf(\mathbf{DI}(e \,\&\, T))}{inf(\mathbf{D})}$$

where information is measured by entropy as

$$inf(\mathbf{D}) = - \sum_i P(C_i) \, log \, P(C_i)$$

$$inf(\mathbf{DI}(e \,\&\, T)) = - \sum_i P(C_i/e \,\&\, T) \, log \, P(C_i/e \,\&\, T)$$

and where the C_i are the constituents of $L(\lambda)$; and

$$(7.12) \quad syst_2 (T \,\&\, e, \mathbf{D}) = \frac{cont(\mathbf{D}) - cont(\mathbf{DI}(T \,\&\, e))}{cont(\mathbf{D})}$$

$$= \frac{1 - \sum_i P(C_i)^2 - P(e \,\&\, T)(1 - \sum_i P(C_i/e \,\&\, T)^2)}{1 - \sum_i P(C_i)^2}$$

$$(7.13) \quad syst_3 (T \,\&\, e, \mathbf{D}) = \frac{cont(\mathbf{D}) - cont(\mathbf{D}/e \,\&\, T)}{cont(\mathbf{D})}$$

$$= \frac{\sum_i P(C_i/e \,\&\, T)^2 - \sum_i P(C_i)^2}{1 - \sum_i P(C_i)^2}.$$

Then, for two theories T_1 and T_2 in $L = L(\lambda \cup \{P\})$

$$(7.14) \quad syst_1 (T_1 \,\&\, e, \mathbf{D}) < syst_1 (T_2 \,\&\, e, \mathbf{D})$$
$$\text{iff } \sum_i P(C_i/e \,\&\, T_1) \, log \, P(C_i/e \,\&\, T_1) <$$
$$< \sum_i P(C_i/e \,\&\, T_2) \, log \, P(C_i/e \,\&\, T_2)$$

(7.15) $syst_2(T_1 \& e, \mathbf{D}) < syst_2(T_2 \& e, \mathbf{D})$
   iff $P(e \& T_1)[1 - \sum_i P(C_i/e \& T_1)^2] >$
$$> P(e \& T_2)[1 - \sum_i P(C_i/e \& T_2)^2]$$

(7.16) $syst_3(T_1 \& e, \mathbf{D}) < syst_3(T_2 \& e, \mathbf{D})$
   iff $\sum_i P(C_i/e \& T_1)^2 < \sum_i P(C_i/e \& T_2)^2.$

Generally, $syst_i(T_1 \& e, \mathbf{D}) < syst_i(T_2 \& e, \mathbf{D})$ holds, for two theories T_1 and T_2 in L, if T_2 is stronger than T_1 in the sense that it excludes more Ct^r-predicates of L than T_1. Note that here again $syst_2$ takes into account the boldness-aspect, whereas $syst_1$ and $syst_3$ are concerned only with likelihoods. Also note that

(7.17)   If $P(C_i/e \& T)=1$ for some i and $P(C_j/e \& T)=0$ for $j \neq i$, then $syst_i(T \& e, \mathbf{D})=1$, for $i=1, 2, 3$.

It should be emphasized that in this chapter, as elsewhere in this work, the extent to which our probabilistic measures of explanatory power capture intuitively valid inductive explanations depends partly on the nature and interpretation of the probability measure used.

In Section 4, we discuss in more detail our measures of systematic or explanatory power, and apply them to the monadic situation by means of Hintikka's inductive logic and our earlier results.

It is easy to see that our measures $syst_i$ and $syst'_i$, $i=1, 2, 3$, have different values when applied to a local explanandum and to a 'corresponding' global explanandum. For example, if our local explanandum is g and the corresponding global explanandum is $\mathbf{D}=\{g, \sim g\}$, then usually $syst_i(T \& e, g) \neq syst_i(T \& e, \mathbf{D})$, and $syst'_i(T \& e, g) \neq syst'_i(T \& e, \mathbf{D})$ for all $i=1, 2, 3$.

In some methodological contexts, it may be of interest to consider even more general measures of explanatory power. Suppose we are working within a research programme within which initially we have a set $\mathbf{H}= \{h_1, \cdots, h_n\}$ of competing hypotheses or theories. Assume that these rivals are mutually exclusive and jointly inductively exhaustive. Let again $\mathbf{D}$ be our domain of basic observational laws (constituents) to be explained. Prior to empirical investigation or at the early stages of such investigation, we do not have much reliable knowledge as to the truth and falsity of the members of **H** or of **D**. We may still try to evaluate the 'worthiness' or the explanatory adequacy of **H**, and compare it with that of some other research programme **H'**, as follows. We ask how much information **H** conveys concerning **D** (possibly given certain initial evidence *e*). This transmitted information (reduction in uncertainty) can be measured by *inf*-entropy or by *cont*-entropy. Analogously with our earlier developments such measures of transmitted information can be normalized to give certain kind of measures $syst_i(\mathbf{H}, \mathbf{D}, e)$ of the (initial) explanatory power of **H** with respect to **D**, given *e*. However, such measures will not be discussed and elaborated in the present work, except for a further brief mention of them in Section 3.

2. INDUCTIVE EXPLANATION ILLUSTRATED

So far, very little has been said about the purely philosophical aspects of explanation. Moreover, the applicability of our ideas to cases of scientific interest has not been discussed. To comment on these aspects of explannation, let us again consider a nonprobabilistic law *g* (in $L(\lambda)$). Why do the observable objects behave so that they satisfy *g*? One way to answer this explanatory question is to introduce a theory *T* (in $L(\lambda \cup \mu)$) from which (possibly together with some initial conditions and rules of correspondence) *g* can be logically *deduced*. Then we say that *T* (potentially) explains *g*.[3]

However, a great many explanations of laws are not clear-cut *deductive* explanations of the above kind. Rather they are *non-deductive* (or inductive). At least many ('common sense') explanations in the social sciences are of this type. For instance, assume that *g* says: All persons of kind *K* (e.g. *K*=schizotymic) exhibit manifest aggression when frustrated. The explanatory theory *T* could be Freud's theory of neuroses. However, *T* does not deductively imply *g* but it only lends some *inductive* support to it, unless the observational property *K* is connected to *T* by means of some correspondence rules. What we have here is only an *explanation sketch* (in Hempel's (1965a) terminology) which, being nondeductive, gives only a *partial* explanation of *g*.

This represents one, but not the only way of looking at the present situation of inductive explanation. For one may as well argue that such an inductive explanation is or may be *epistemically complete*, i.e., complete with respect to all the information we have at our disposal. Furthermore one may in some situations argue even that no more information is obtainable, and that what we now have adequately represents reality.

We will not here take any definite standpoint on the various ways of subjecting inductive explanation to philosophical interpretation as our methodological results seem to comply with various views. Instead, we present some examples to highlight our abstract discussion in this chapter.

It would be difficult to give an exhaustive philosophical classification of types of inductive explanation, and only some of them are mentioned here. First, we have the inductive explanation of observable *dispositions* (analysed in terms of observational generalizations). For instance, the fact that an object is soluble in water means that something like the following lawful statement holds true: In all cases when the object becomes immersed in water under such-and-such conditions, the object dissolves. At least for a philosophical 'actualist' this kind of disposition-law represents regularities that call for a deeper explanation. An actualist would look for some hidden (structural) properties of the object, in order to find an explanation of the disposition-law. Most often such an explanation is *inductive* (only). Indeed, one may argue, had we a deductive explanation in terms of some deeper chemico-physical theoretical properties, there would no longer be any need for disposition-talk.

The phosphorus-example of Hempel discussed in Chapter 1.1, can be regarded as representing an inductive explanation of disposition-laws. A technical treatment of this example was given in Example 3.1. There, we noted the central roles in this explanation played by the theoretical concept of phosphorus, and by the idea of positive inductive relevance (and hence most of our above measures) as measuring explanatory strength.

As a second type of inductive explanation we wish to mention the *How possible?* explanations common in such disciplines as history and the social sciences. We may classify explanations according to whether they give sufficient or necessary conditions of the explanandum. The former answer questions of the form *Why necessary?* while the latter are answers to *How possible?*. It seems to us that inductive explanation in terms of positive relevance accounts for some central logical aspects of the latter type of explanation. For if a statement F is explained by citing a contingent proposition G necessary for F (i.e., $F \vdash G$), then obviously $P(F/G) >$

$>P(F)$. Explanations in archaeology are often of the *How possible?* type. Reference can be made to Heyerdahl's (1971) explanation concerning cross-cultural archaeological similarities. This explanation relies on his reed boat crossing the Atlantic. This represents an attempt to support archaeological diffusionism against isolationism.

Thirdly, it seems that at least some *teleological* or purposive explanations in biology (and possibly elsewhere) are best construed as inductive explanations in our sense. To give an example, let g be an observational generalization saying that animals of a certain kind (say O_1) have kidneys (O_2). Now we introduce an explanatory theory T: The function of kidneys is to remove waste chemicals and to regulate the amount of water in the tissues, which is necessary for the proper working of the organism.

To make matters simple, let 'O_3' stand for the complex observational property of having no waste chemicals and being in a normal state as to the amount of water in tissues, and 'P' for the theoretical property of being in the proper working condition. Thus we have $\lambda=\{O_1, O_2, O_3\}$ and $\mu=\{P\}$ and, within a monadic framework, we suggest the following formalization:

$$g = (x)\,(O_1\,(x) \supset O_2\,(x))$$
$$T = (x)\,(P\,(x) \supset O_3\,(x))\ \&\ (x)\,(O_2\,(x) \supset O_3\,(x))$$

Let the biologist's empirical evidence be given by e. Then clearly *not* $(T\ \&\ e) \vdash g$. Still, as our earlier technical results indicate, it is possible to have $P(g/T\ \&\ e)>P(g)$ and thus an inductive explanation of g by means of T, given e.

Finally, we are faced with the broad spectrum of cases when only lack of information forces us to accept inductive explanations instead of deductive ones. (One may of course argue that this is the only reason in the above purposive explanation and in the earlier explanations of dispositions.) Most often in these cases we miss correspondence rules connecting the theoretical concepts of a theory T and the generalization to which T is extended to apply,[4] or the correspondence rules may be too hazy and vague to be explicitly and fully stated. Good examples of cases of the latter type are to be found in personality psychology. The potential psychoanalytic explanation mentioned earlier is one of these. Sherwood (1969) gives a large number of other such examples taken from psychoanalytic theorizing. He rejects the ordinary deductive account of explanation for psychoanalytic cases. Instead he suggests that individual cases be explained by means of what he terms a *psychoanalytic narrative*. What is interesting from our point of view is that such a narrative, as its elements, normally contains observational generalizations and their inductive (or non-deductive) explanations by means of Freud's theory (cf. in particular Chapter 7 of Sherwood, 1969).

Consider for instance the following empirical generalization established by Freud and discussed by Sherwood (1969), p. 212ff.: All young boys between the ages of three and five pass through a period of intensive rivalry with and hostility toward their fathers over the attentions of their mothers. This generalization loudly cries for explanation. Within Freud's theory, a (potential!) inductive explanation of this can be given by means of the theoretical concept of Oedipal complex, which is one of the *causal* factors involved. Improvement in knowledge may result in stronger and stronger causal explanations perhaps approaching deductive explanation. To conclude, we may cite Sherwood on the use such artificial limitations as 'between the ages of three and five' in empirical generalizations: "Usually these limitations are meant to be interpreted as implicative, shorthand notations for other, unmentioned causal factors, *even though such factors may not yet be known.* In such cases, there is only the implied *conviction* that causal factors can be discovered to account for the peculiarities of behaviour which characterize the groups singled out by the rather arbitrary categories employed." (Sherwood (1969), p. 212, original italics).

3. POSITIVE INDUCTIVE RELEVANCE, SUPERSESSANCE, AND SCREENING OFF

3.1.   When does a theory *supersede* another theory in (inductively) explaining a certain law, set of laws, or some other explanandum? An adequate answer to this question naturally presupposes that one has a clear view of (inductive) explanation. The notion of explanation is notoriously one of the most difficult philosophical concepts to elucidate, as it has a pragmatic flavour. It seems that in a fully adequate account of explanation one would have to take into account, among other things, the socio-historical context, as well as some psychological aspects of the explainer and the explained. However, as was the case earlier in this chapter, we will try to abstract from such pragmatic aspects and think of them as fixed boundary conditions. Consequently, below we rely upon the methodological analysis of inductive explanation given in the last section.

The first elucidation of the notion of supersessance that Section 1 suggests is this. A theory is said to supersede another theory in explaining a given law exactly when the explanatory or systematic power of the first theory with respect to the law exceeds that of the latter theory. However, as we have several different measures of explanatory power we obtain as many different notions of explanatory supersessance. Thus, with a view to limiting our analysis, we concentrate on those measures which satisfy the likelihood condition (7.9) of Section 1. That is, we have chosen to elucidate explanatory supersessance by means of *positive inductive relevance* as follows: A theory T_2 *supersedes* a theory T_1 *in explaining* a generalization g, given the evidence e, if and only if

(7.18) $P(g/T_2 \& e) > P(g/T_1 \& e) > P(g/e)$.

In this definition T_1 may be the null theory in $L(\lambda \cup \mu)$, viz. a theory earlier characterized by $r=0$ and $b'=2b$ within our system. Note that the probability $P(g/e)$ is defined for $L(\lambda)$, as g and e are normally taken to be observational or empirical, while the other probabilities $P(g/T_1 \& e)$ and $P(g/T_2 \& e)$ are of course defined for $L(\lambda \cup \mu)$.

Of our measures of systematic or explanatory power defined in Section 1, $syst_i$ and $syst'_1$ satisfy the above likelihood condition when $i=1, 3$. It is worth recalling here that the measures $syst_2$ and $syst'_2$ require both high positive relevance *and* boldness (with respect to background information) of a good explanatory theory. Hence, the results we intend to establish for our notion of supersedence characterize explanation as elucidated by $syst_2$ and $syst'_2$ only in the case of equally bold theories. (See our discussion in Section 4.)

In the above definition of supersessance we may replace g by **D** (the set of constituents of $L(\lambda)$) to acquire a more global notion of explanatory supersessance. (However, on the right hand side of our definition we have to use one of the conditions (7.14)–(7.16).) The resulting explanandum-global notion of supersessance might be taken to explicate the (*inductive*) *reduction* of T_1 into T_2. This notion of reduction (on the basis of the observational statements explained) is thus our inductive counterpart to the Kemeny-Oppenheim-type deductive reduction (see Kemeny and Oppenheim, 1956).

Our above notion of inductive reduction is indirect and incomplete in the sense that nothing direct is stated about the relationship between the basic assumptions of T_1 and those of T_2. As an improvement, we suggest:

> T_2 *reduces* T_1, given e, if and only if
> (a) T_2 explains T_1, given e, and
> (b) T_2 supersedes T_1 in explaining **D**, given e.

This definition obviously covers both deductive and inductive reduction, but we are here only interested in the inductive explicates of the key notions. Thus (a) might be explicated by either

(a′) $P(T_1/T_2 \ \& \ e) > P(T_1/e)$ or by

(a″) $P(T_1/T_2 \ \& \ e) - P(T_1/e) > P(T_2/T_1 \ \& \ e) - P(T_2/e) > 0.$

It can easily be verified that (a″) equals (a′) *and* the requirement $cont(T_2/e) > cont(T_1/e)$. This latter condition is the boldness condition discussed in Section 1.

If condition (b) is omitted from our definition, we have a weakened notion of (inductive) reduction which corresponds to Nagel's intertheoretic (deductive) reduction (see Nagel, 1961).

In still another important sense, a theory T_2 may be said to be a better explanans of g than is a theory T_1. In this sense, T_2 is better than T_1 exactly when the introduction of T_2 makes T_1 *inductively irrelevant* to g in the presence of T_2. This is the notion (or actually the notions) of *screening off* advocated by Salmon (1971a) and by Greeno (1971). We first discuss our counterpart (call it screening$_1$ off) to Salmon's definition of this notion in Salmon (1971a), p. 55. Our screening$_1$ off is formally the same as Salmon's the only essential difference being that Salmon uses a set theoretical framework as distinct from our 'linguistic' framework with general statements. Following Salmon (1971a), p. 55, we define:

(7.19)   A theory T_2 *screens*$_1$ *off* a theory T_1 from a generalization g, given the evidence e, if and only if
$$P(g/T_1 \ \& \ T_2 \ \& \ e) = P(g/T_2 \ \& \ e) \neq P(g/T_1 \ \& \ e).$$

In this definition the function of the inequality is mainly that of making the notion of screening$_1$ off non-symmetric (with respect to its first (T_2)

and second (T_1) arguments), a feature which distinguishes it from the symmetric notion of (relative) inductive irrelevance.

3.2.   Let us now investigate the notions of supersessance and screening$_1$ off. We may start by comparing the inductive probabilities $P(g/T_1 \& e)$ and $P(g/T_2 \& e)$ in general. By the general multiplication theorem of probability calculus, we have for the present case

$$(7.20) \quad P(g/T_1 \& T_2 \& e) = \frac{P(g/T_1 \& e)\, P(T_2/g \& T_1 \& e)}{P(T_2/T_1 \& e)}$$
$$= \frac{P(g/T_2 \& e)\, P(T_1/g \& T_2 \& e)}{P(T_1/T_2 \& e)}.$$

On the basis of (7.20), we immediately derive an expression for the comparison of $P(g/T_1 \& e)$ and $P(g/T_2 \& e)$:

$$(7.21) \quad \frac{P(g/T_2 \& e)}{P(g/T_1 \& e)} = \frac{P(T_1/T_2 \& e)\, P(T_2/g \& T_1 \& e)}{P(T_2/T_1 \& e)\, P(T_1/g \& T_2 \& e)}$$
$$= \frac{P(T_1 \& e)\, P(T_2/g \& T_1 \& e)}{P(T_2 \& e)\, P(T_1/g \& T_2 \& e)}.$$

Thus we also have

(7.22)   When $P(g/T_1 \& e) > P(g/e)$ and $P(g/T_2 \& e) > P(g/e)$,
T_2 supersedes T_1 in explaining g, given e, iff
$P(T_1/e)\, P(T_2/g \& T_1 \& e) > P(T_2/e)\, P(T_1/g \& T_2 \& e)$
iff
$P(T_1/e)\, P(T_2/g \& e) > P(T_2/e)\, P(T_1/g \& e)$.

As we see from (7.22) our notion of *supersessance* (7.18) also takes into account the boldness-factor, although only in a relative sense. Note also that no close connection (i.e., necessary or sufficient conditionship) exists between supersessance and the reduction-explicates (a') or (a''), which is natural. (Of course, the connection with the reduction condition (b) is close; it is the identity relation if $\mathbf{D} = \{g\}$.)

Consider now the special case $T_2 \vdash T_1$. Does the substitution of T_1 by a deductively stronger theory T_2 entail that T_2 supersedes T_1 in explaining g (given that both of them explain g severally)? If $T_2 \vdash T_1$, then $P(T_1/T_2 \& e) = 1$, $P(T_1/g \& T_2 \& e) = 1$. Hence we have by (7.21)

(7.23)   If $T_2 \vdash T_1$, then

$$P(g/T_1 \& T_2 \& e) = P(g/T_2 \& e)$$
$$= \frac{P(T_2/g \& T_1 \& e)}{P(T_2/T_1 \& e)} P(g/T_1 \& e).$$

Hence here

$$P(g/T_2 \& e) > P(g/T_1 \& e) \quad \text{iff}$$
$$P(T_2/g \& T_1 \& e) > P(T_2/T_1 \& e) \quad \text{iff}$$
$$P(g/T_1 \& T_2 \& e) > P(g/T_1 \& e).$$

It can thus be seen that the subtheory T_1 of T_2 is not necessarily an inferior explanans (in the positive relevance and supersessance sense) of g in comparison with T_2. However, T_2 will prove to be superior as soon as it is positively relevant to g in the presence of T_1.

As to screening₁ off, in the present case we obtain:

(7.24)   If T_1 is a subtheory of T_2 (i.e., $T_2 \vdash T_1$) then $P(g/T_1 \& T_2 \& e)$ $=P(g/T_2 \& e)$, for any g and e.

If, furthermore, $P(g/T_2 \& e) \neq P(g/T_1 \& e)$ in this case, T_2 screens₁ off T_1 from g, given e. The assumption $P(g/T_1 \& T_2 \& e)=P(g/T_2 \& e)$ does not imply that $P(g/T_1 \& e) \neq P(g/T_2 \& e)$ even when $T_2 \vdash T_1$, and not $T_1 \vdash T_2$, as can be seen from (7.23) and (7.24). Thus, we have the result

(7.25)   If T_1 is a subtheory of T_2, then T_2 screens₁ off T_1 from g, given e, only if T_2 is (positively or negatively) relevant to g relative to $T_1 \& e$.

From (7.20) and the definitions of screening₁ off and supersessance we obtain the following interesting results:

(7.26)   If T_2 screens₁ off T_1 from g, given e, then it need not be the case that T_2 supersedes T_1 in explaining g, given e.

(7.27)   If T_2 supersedes T_1 in explaining g, given e, then it need not be the case that T_2 screens₁ off T_1 from g, given e.

(7.28)   If T_2 screens₁ off T_1 from g, given e, and T_1 is positively relevant to g, relative to e, then it need not be the case that T_2 is positively relevant to g, relative to e.

Our negative results (7.26) and (7.28) can be illustrated by a simple numerical example. Suppose that $T_2 \vdash T_1$ and that $P(T_1/e)=0.50$, $P(T_2/e)=0.40$, $P(g/e)=0.27$, $P(T_1 \ \& \ g/e)=0.15$, and $P(T_2 \ \& \ g/e)=0.10$. Then $P(g/T_1 \ \& \ e)=0.30$ and $P(g/T_2 \ \& \ e)=0.25$, whence

$$P(g/T_1 \ \& \ e) > P(g/e) > P(g/T_2 \ \& \ e),$$

even if T_2 screens$_1$ off T_1 from g, given e.

It is easily demonstrable that

(7.29)   If T_2 screens$_1$ off T_1 from g, given e, and if $\sim T_1 \ \& \ T_2 \ \& \ e$ is consistent, then T_2 screens$_1$ off also $\sim T_1$ from g, given e.

It should, however, be noted that if $P(g/T_1 \ \& \ T_2 \ \& \ e)=P(g/T_2 \ \& \ e)$, then in general $P(g/T_1 \ \& \ \sim T_2 \ \& \ e) \neq P(g/\sim T_2 \ \& \ e)$. This result is obtained direct from (7.20) which gives

$$P(g/T_1 \ \& \ \sim T_2 \ \& \ e) =$$
$$= \frac{P(g/T_1 \ \& \ e) - P(g/T_2 \ \& \ e) \, P(T_2/T_1 \ \& \ e)}{1 - P(T_2/T_1 \ \& \ e)},$$

which usually differs from

$$P(g/\sim T_2 \ \& \ e) = \frac{P(g/e) - P(g/T_2 \ \& \ e) \, P(T_2/e)}{1 - P(T_2/e)}.$$

This means that if T_2 screens$_1$ off T_1 from g, given e, then it need not be the case that $\sim T_2$ screens$_1$ off T_1 from g, given e.

The upshot of our discussion concerning the relationships between the notions of inductive supersessance and screening$_1$ off comes to this, that neither of these concepts implies the other one. Furthermore, the analogous statement holds true for the relationship between both of these concepts and our notion of intertheoretic reduction (in sense (a) of p. 98).

As such, the fact that screening$_1$ off and supersessance are so different notions may not seem bad. However, we claim that screening$_1$ off is a methodologically uninteresting notion, partly because of its failing to share some important properties with supersessance. To repeat, screening$_1$ off has two main faults (cf. our results (7.24), (7.25), and (7.28)). First, suppose T_1 is positively relevant to g, given e, and hence T_1 explains g.

Now, if some theory T_2 screens$_1$ off T_1, T_2 need not be positively relevant to g and explain g. Of course, screening$_1$ off cannot then be used to measure explanatory betterness or supersessance. The second main methodological disadvantage with the notion of screening$_1$ off is this: it is generally too easy to find a theory T_2 which screens$_1$ off a T_1 from g. Namely, every T_2 such that $T_2 \vdash T_1$ and $P(g/T_1 \& e) \neq P(g/T_2 \& e)$ would do. For instance, T_1 might be the Craigian transcription (in $L(\lambda)$) of T_2. However, such a stronger theory T_2 might as well be just an artifact. What is more, we may have $P(g/T_2 \& e) < P(g/T_1 \& e)$ for such a T_2 so that T_2 does not supersede T_1.

3.3.   Let us next turn to our counterpart of Greeno's notion of screening off (see Greeno, 1970, 1971). Like Salmon, Greeno employs a set-theoretical framework. Greeno's notion of screening off can be formally translated into our framework for explaining generalizations by means of theories. However, it is argued that the methodological importance of this notion is not very great.

We now consider a methodological situation in which we have a set $\mathbf{H} = \{h_1, ..., h_n\}$ of competing hypotheses or theories for explaining a set $\mathbf{D}$ of observational statements. This is exactly the situation we consider at the very end of Section 1. We may here also apply the same methodological interpretation as there, viz. the initial evaluation of the explanatory adequacy of competing research programmes $\mathbf{H}$, $\mathbf{H}'$, ...

For simplicity, we normally assume below that our sets of hypotheses contain only two members: a hypothesis or theory and its negation. Let us thus consider the sets $\mathbf{T}_1 = \{T_1^1, T_1^2\}$, $\mathbf{T}_2 = \{T_2^1, T_2^2\}$ and $\mathbf{g} = \{g^1, g^2\}$. Here $T_1^1 = T_1$ and $T_1^2 = \sim T_1$, and the analogous notation also applies to the sets $\mathbf{T}_2$ and $\mathbf{g}$, too. Reference to evidence e will be omitted to avoid notational complexity.

Next we consider the *entropy transmitted* from one set of hypotheses $(\mathbf{T}_1)$ to another $(\mathbf{g})$. 'E' being used for logarithmic entropy, and 'V' for transmitted entropy, this well known measure becomes (cf. Greeno, 1970):

$$(7.30) \quad V(\mathbf{T}_1, \mathbf{g}) = E(\mathbf{T}_1) + E(\mathbf{g}) - E(\mathbf{T}_1 \cap \mathbf{g}).$$

The logarithmic entropies in (7.30) are defined as follows in analogy with the general finite case:

$$(7.31) \quad E(\mathbf{T}_1) = \sum_{i=1}^{2} - P(T_1^i) \, log \, P(T_1^i)$$

(7.32) $E(\mathbf{g}) = \sum\limits_{j=1}^{2} - P(g^j) \log P(g^j)$

(7.33) $E(\mathbf{T_1} \cap \mathbf{g}) = \sum\limits_{j=1}^{2} \sum\limits_{i=1}^{2} - P(T_1^i) P(g^j/T_1^i) \log P(T_1^i) P(g^j/T_1).$

Similarly, we have the definition for $V(\mathbf{T_2}, g)$, the entropy transmitted from set $\mathbf{T_2}$ to set $\mathbf{g}$. Analogously with Greeno (1970), one may use the unnormed transmitted entropy for direct measurement of the explanatory power of a set of theories concerning a set of observational statements. As we said in Section 1, our various measures of explanatory power can easily be extended to the present case, as they are simply normed measures of transmitted information.

In any case, either $V(\mathbf{T_1}, \mathbf{g})$ itself or one of its normed versions may be taken as a measure of the reduction of the uncertainty associated with $\mathbf{g}$, which reduction is due to $\mathbf{T_1}$. The question then arises whether one can improve on the explanation provided by $\mathbf{T_1}$ by introducing a better 'theory' (or set of hypotheses). Let $\mathbf{T_2}$ be such a candidate. We can then say that $\mathbf{T_2}$ is a better explanans of $\mathbf{g}$ than $\mathbf{T_1}$ if and only if $\mathbf{T_2}$ *screens$_2$* *off* T_1 from $\mathbf{g}$ in the sense that, for all i, j, k,

(7.34) $P(g^j/T_1^i \ \& \ T_2^k) = P(g^j/T_2^k).$

Here, it has to be assumed that all the conjunctions $T_1^i \ \& \ T_2^k$, $i, k = 1, 2$, are consistent. In our special case of two-member sets, we can replace (7.34) simply by

(7.35) $P(g/T_1 \ \& \ T_2) = P(g/T_2)$
$P(g/T_1 \ \& \sim T_2) = P(g/\sim T_2).$

From (7.35), it is immediately discernible that screening$_2$ off is a stronger notion than screening$_1$ off. In fact, we obtain the results

(7.36)   If T_2 screens$_1$ off T_1 from g and if $\sim T_2$ screens$_1$ off T_1 from g, then $\mathbf{T_2} = \{T_2, \sim T_2\}$ screens$_2$ off $\mathbf{T_1} = \{T_1, \sim T_1\}$ from $\mathbf{g} = \{g, \sim g\}$.

(7.37)   If $\mathbf{T_2} = \{T_2, \sim T_2\}$ screens$_2$ off $\mathbf{T_1} = \{T_1, \sim T_1\}$ from $\mathbf{g} = \{g, \sim g\}$, and if $P(g/T_2) \neq P(g/T_1)$ and $P(g/\sim T_2) \neq P(g/T_1)$, then T_2 screens$_1$ off T_1 from g and $\sim T_2$ screens$_1$ off T_1 from g.

Let $\mathbf{T}_{12} = \{T_1^i \,\&\, T_2^k \mid i, k = 1, 2\}$. The screening$_2$ off condition (7.35) implies

$$
\begin{aligned}
P(g^j/T_1^i) &= \sum_{k=1}^{2} P(T_2^k/T_1^k)\, P(g^j/T_1^i \,\&\, T_2^k) \\
&= \sum_{k=1}^{2} P(T_2^k/T_1^i)\, P(g^j/T_2^k).
\end{aligned}
$$

It follows that $V(\mathbf{T}_{12}, g) = V(\mathbf{T}_2, g)$ when the screening$_2$ off condition is satisfied. On the other hand, we have $V(\mathbf{T}_{12}, \mathbf{g}) \geqslant V(\mathbf{T}_1, \mathbf{g})$. Consequently, the basic result which holds for the present screening$_2$ off notion is this: (cf. Greeno, 1970, pp. 289–291 for a proof within a set-theoretical framework):

(7.38)   If $\mathbf{T}_2$ screens$_2$ off $\mathbf{T}_1$ from $\mathbf{g}$ then $V(\mathbf{T}_2, \mathbf{g}) \geqslant V(\mathbf{T}_1, \mathbf{g})$.

Thus, $\mathbf{T}_2$'s screening$_2$ off $\mathbf{T}_1$ automatically entails that no loss in explanatory power occurs, provided explanatory power covaries with transmitted entropy.

Our main criticism of the notion of screening$_1$ off was that it does not have a proper connection with the notion of positive relevance (cf. result (7.28)). In this respect, the notion of screening$_2$ off fares better than that of screening$_1$ off. To see this, suppose that T_1 is positively relevant to g, and that $\mathbf{T}_2$ screens$_2$ off $\mathbf{T}_1$ from $\mathbf{g}$. If now T_2 is negatively relevant to g, then $\sim T_2$ is positively relevant to g. If T_2 is irrelevant to g, so that $P(g/T_2) = P(g/\sim T_2) = P(g)$, then, by the application of (7.35),

$$
\frac{P(g \,\&\, T_1/T_2)}{P(g \,\&\, T_1/\sim T_2)} = \frac{P(T_1/T_2)}{P(T_1/\sim T_2)}.
$$

This equality is reduced to

$$
P(g/T_1 \,\&\, T_2) = P(g/T_1),
$$

from which, by the screening$_2$ off and irrelevance assumptions,

$$
P(g/T_1) = P(g/T_1 \,\&\, T_2) = P(g/T_2) = P(g).
$$

This implies that T_1 is irrelevant to g, which contradicts the assumption $P(g/T_1) > P(g)$. We thus have the following result for two-member sets $\mathbf{T}_1, \mathbf{T}_2$, and $\mathbf{g}$:

(7.39)   If T_2 screens$_2$ off T_1 from g, and if one of the elements of T_1
is positively relevant to g, then one of the elements of T_2 is
positively relevant to g.

However, this proof does not hold true for sets T_1, T_2, and g with more
than two members.

To begin our critical comments on the notion of screening$_2$ off, it needs
to be pointed out that it is a very strong notion. In fact, we claim that
while screening$_1$ off was found to be too weak a notion, screening$_2$ off
seems to be too strong. Condition (7.35) requires that *both* T_2 and $\sim T_2$
make T_1 irrelevant to g. It is hard to find methodologically interesting
situations in which this would be the case. In the first place, if T_2 has a
great information content, then $\sim T_2$ has a small one. However, (7.35)
implies that T_2 and $\sim T_2$ play here a symmetrical role. These two facts
seem hard to reconcile. Secondly, it is easy to see that any T_2 and its
negation will satisfy (7.35) in the event that T_1 is tautology. However, if
T_1 is a strong or informative theory, it seems unlikely that any methodologically interesting theory T_2 which satisfies the first clause of (7.35) will
also satisfy its second clause. Thirdly, condition (7.34) involves the assumption that all the conjunctions

$$T_1 \& T_2$$
$$\sim T_1 \& T_2$$
$$T_1 \& \sim T_2$$
$$\sim T_1 \& \sim T_2$$

are consistent. This means, for instance, that T_1 cannot be a subtheory
of T_2 or of $\sim T_2$. More generally, there cannot be any deductive connections between the members of T_1 and T_2 (cf. below our criticism of
this situation within the set-theoretic framework). Notice that this compatibility requirement excludes both cumulative growth (by means of extensions of theories) and 'Feyerabendian growth' by successive incompatible theories from possible models for scientific change.

3.4.   On the basis of this discussion, we can now make some general
comments on Greeno's and Salmon's ideas of inductive and statistical
explanation.

One way of viewing the differences between explanatory power as measured by positive relevance (or by measures, such as $syst_1(T \& e, g)$ and $syst_3(T \& e, g)$, directly proportional to it) as opposed to transmitted entropy is this that transmitted entropy explicitly deals with rivalling theories or, as in our special case, with negations of theories, whereas our explanandum-local positive relevance measures of Section 1 do this only implicitly. However, this is only one side of the coin, for the issue seems to become a matter of different intuitions about inductive explanation. This is reflected, for instance, in the minimum values of the two types of measures of explanatory power. Transmitted entropy-measures for sets of explanantia and explananda, such as Greeno's, obtain their minimum value when the explanans and the explanandum are inductively independent, i.e., when there is no reduction in the uncertainty associated with the explanandum due to the explanans. On the other hand, such measures as our $syst_1(T \& e, g)$ and $syst_3(T \& e, g)$ receive their minimum values, not for the independence case but when $T \& e \vdash \sim g$. From the viewpoint of transmitted entropy, however, the case $T \& e \vdash \sim g$ means on the contrary that no uncertainty remains in regard to the set $\mathbf{g} = \{g, \sim g\}$, given $T \& e$.[5]

The above remarks indicate that an important difference exists between conceptions of inductive explanation which allow negative relevance to have explanatory power and those which do not allow it.

Salmon's statistical relevance model (S-R model) for the probabilistic explanation of particular events or facts is based on the following idea (cf. Salmon, 1971a, pp. 76–77). To explain why a member x of a class A has the property B, we partition A into subclasses $A \cap C_1$, $A \cap C_2, \ldots,$ $A \cap C_n$, all of which are homogeneous with respect to B, we give the probabilities $P(B/A \cap C_k)$, all of which are different from each other, and we say into which class $A \cap C_i$ x belongs. The requirement of homogeneity of the subclasses means, in effect, that each C_i, $i = 1, \ldots, k$, screens$_1$ off every property D (or, alternatively, every property D known or believed to be a relevant factor) from B, given A. This amounts to requiring that the set $\{C_1, \ldots, C_k\}$ screens$_2$ off the set $\{D, -D\}$, for each property D, from the set $\{B, -B\}$, given A. Therefore, our earlier criticism of the notions of screening$_1$ off and screening$_2$ off are applicable here, despite Salmon's framework being set-theoretical.

Even though Salmon (1971a, p. 42) claims that the notion of statistical (inductive) relevance is the fundamental concept upon which he hopes to build his explication of inductive explanation, his model allows of explanations in which the explaining property C_i is positively relevant, negatively relevant or even irrelevant to the explanandum (cf. Salmon, 1971a, pp. 62–65). Consequently, Salmon is forced to abandon his basic intuition that the explanans should be statistically relevant to the explanandum (see, e.g., Salmon, 1971a, p. 11). In view of our results, this is because he wants to base the S-R model merely upon the idea of screening off, even if this notion does not have an intimate connection with the notions of positive and negative relevance. As we have emphasized in this work the methodological importance of the notion of positive relevance, particularly in the context of theoretical explanation of laws, the notion of screening off plays no important role within our account.

Salmon explicitly restricts the applicability of his model to the explanation of particular facts and admits that theoretical explanation (i.e., explanation of laws by means of theories) may have a logical structure entirely different from that explicated in the S-R model (cf. Salmon, 1971a, p. 81). As we have argued, the explanation of laws by applying theories introducing theoretical concepts is a central task of science. In contrast to Salmon, our approach to inductive explanation is concerned with the theoretical explanation of laws. Our account can be considered anti-inductivist and anti-atomistic. For we explicitly take into account theoretical background information which specifies the factual and analytic connections obtaining within the conceptual framework employed. This is in clear contrast with Salmon's and Greeno's approaches. They are only concerned with sets or random variables among which no deductive connections are assumed to exist initially. (Cf. the compatibility requirement pertaining to (7.34).) Salmon does not employ theoretical concepts at all, and Greeno's theoretical concepts can be regarded as merely instrumentalistic intervening variables (cf. also Jeffrey, 1971, pp. 42–43 on this problem).

4. INDUCTIVE EXPLANATION WITHIN HINTIKKA'S SYSTEM

4.1.   The preceding sections introduced and discussed some information-theoretic measures of the systematic or explanatory power of theories with respect to observational generalizations. We continue with some results obtained in regard to these measures within Hintikka's system of inductive logic. The results and examples of this section illustrate the role that theoretical terms can play within the kind of inductive explanation discussed above.

A beginning is made with our measures of systematic power that are local with respect to the explanandum. Accordingly, we let g be a generalization of $L(\lambda)$, and compare the values of the explanatory power of two rivalling theories T_1 and T_2 with respect to g, given a piece of evidence e. Here the theories T_1 and T_2 may both be in the same language or they may be in different languages. The most interesting cases from a methodological aspect are the following two:

(a) T_1 and T_2 are both in $L(\lambda \cup \{P\})$
(b) T_1 is in $L(\lambda \cup \{P\})$, but T_2 is in $L(\lambda)$.

We first consider Case (a). Thus, we assume that T_1 and T_2 are two theories in $L(\lambda \cup \{P\})$ which explain g inductively, given e, in the positive relevance sense. Then, according to (7.18), T_1 supersedes T_2 in explaining g, given e, if and only if

(7.40) $P(g/e \ \& \ T_1) > P(g/e \ \& \ T_2) > P(g/e)$.

When (7.40) holds, we know by (7.8) that, for the measures $syst_i$ and $syst'_i$, $i = 1, 3$, the systematic or explanatory power of T_1 with respect to g, given e, is greater than that of T_2. We write '$S(T_1, T_2, g, e)$' for 'T_1 supersedes T_2 in explaining g, given e' and give some simple conditions under which this relation holds.

Suppose first that 'P' is an evidential theoretical term. With application of the same symbolism as earlier, with indices '1' and '2' referring to theories T_1 and T_2, we obtain from cases (4), (5), and (6) of Chapter 3.3, for a weak generalization g:

(7.41)   If $\alpha = n$,   then
          $S(T_1, T_2, g, e)$ iff $b'_1 < b'_2 < b$.

(7.42)   If α and n are great, then
          $S(T_1, T_2, g, e)$ iff $b'_1 < b'_2 < b$.

(7.43)   If $b + c = K$, $b = b'_i$, and $b'_i + c'_i + r_i = 2K$, for $i = 1, 2$,
          then $S(T_1, T_2, g, e)$ iff $c'_1 > c'_2$ and $\alpha > n$.

For strong generalizations $(g = C_w)$, the following result is derived:

(7.44) If α and n are great, then

$$S(T_1, T_2, C_w, e) \quad \text{iff}$$

$$\left(2 + \frac{\alpha}{n}\right)^{K - r_i - c_{0i} + b - b_i'} > \left(1 + \frac{\alpha}{n}\right)^{K - r_i - c_{0i}}, \quad i = 1, 2,$$

and

$$\left(2 + \frac{\alpha}{n}\right)^{r_2 + c_{02} + b_2' - r_1 - c_{01} - b_1'} < \left(1 + \frac{\alpha}{n}\right)^{r_2 + c_{02} - r_1 - c_{01}}.$$

(Here $c_{0i} = c_i' - c$, for $i = 1, 2$.) If $b = b_1' = b_2'$, then $S(T_1, T_2, C_w, e)$ holds, for great α and n, if and only if

$$K - r_i - c_{0i} > 0, \quad \text{for} \quad i = 1, 2,$$

and

$$r_2 + c_{02} - r_1 - c_{01} > 0.$$

This last result also holds when $\alpha = n$.

In cases (7.41) and (7.42), the superseding theory T_1 has to be stronger than T_2 in the sense of requiring more empty cells in the 'area' of the empty observational cells corresponding to g (i.e., $b_1' < b_2'$). As there also $b_i' < b$, for $i = 1, 2$, our theories have observational consequences, T_1 being observationally stronger than T_2, at least as far as the 'application field' of g is concerned. This type of condition is thus directly relevant to the explanation of g. In other cases, where our theories need not have observational consequences, we have restrictions concerning the variety of evidence with respect to the richer language L or the logical strength of the theories in terms of r_1 and r_2.

The results (7.41) and (7.42) also hold in the event of 'P' being a non-evidential theoretical term.

One can also consider cases in which 'P' is evidential in T_1, and non-evidential in T_2. We illustrate this by continuing Example 3.1. Let $\lambda = \{O_1, O_2\}$, $\mu = \{P\}$,

$$g = (x) (O_1(x) \supset O_2(x)),$$

and

$$T = (x) [(P(x) \supset O_1(x) \& (P(x) \supset O_2(x))].$$

Let T_1 be the theory T where 'P' is evidential. Similarly, let T_2 be the theory T where 'P' is non-evidential. Suppose further that all cells of $L(\lambda)$ except 'O_1 & $\sim O_2$' are exemplified in our evidence e, and that the size n of sample e is less than α. If now the individuals of e belonging to $Ct_1 = = O_1$ & O_2 split into both Ct_1 & P and Ct_1 & $\sim P$, then

$$P(g/e) < P(g/e \ \& \ T_2) < P(g/e \ \& \ T_1).$$

In other words, T_1 supersedes T_2 in explaining g, given e. However, if the individuals of e belonging to Ct_1 do not split into both subcells, then

$$P(g/e) < P(g/e \ \& \ T_1) < P(g/e \ \& \ T_2),$$

so that here T_2 supersedes T_1 in explaining g, given e. Thus we see that a non-evidential theoretical predicate can lead to a better explanation of an observational generalization than an evidential one if the evidence does not show sufficient variety with respect to the latter.

The definition (7.40) for supersessance is based upon the measures $syst_i$ and $syst_i'$, $i = 1, 3$. However, $syst_2$ and $syst_2'$ lead to the following notion of supersessance (call it supersessance$_2$): If T_1 and T_2 explain g inductively, given e, then T_1 supersedes$_2$ T_2 if and only if

(7.45) $P(T_1/e) [1 - P(g/e \ \& \ T_1)] < P(T_2/e) [1 - P(g/e \ \& \ T_2)]$

(cf. (7.9)). Since, by formulas (2.8') and (2.9),

$$
(7.46) \quad P(T/e) = \frac{\displaystyle\sum_{i=0}^{2K-r-c'} \binom{2K - r - c'}{i} \frac{(\alpha + c' + i - 1)!}{(n + c' + i - 1)!}}{\displaystyle\sum_{i=0}^{2K-c'} \binom{2K - c'}{i} \frac{(\alpha + c' + i - 1)!}{(n + c' + i - 1)!}},
$$

we know that $P(T_1/e) = P(T_2/e)$ whenever $c_1' = c_2'$ and $r_1 = r_2$. Therefore, all of our results quoted above also apply to the notion of supersessance$_2$ whenever T_1 and T_2 are theories such that $c_1' = c_2'$ and $r_1 = r_2$. However, when $P(g/e \ \& \ T_1) = P(g/e \ \& \ T_2)$, so that $syst_i(T_1 \ \& \ e, g) = syst_i(T_2 \ \& \ e, g)$ and $syst_i'(T_1 \ \& \ e, g) = syst_i'(T_2 \ \& \ e, g)$, for $i = 1, 3$, the bolder theory of T_1 and T_2 will supersede$_2$ the other. If $c_1' = c_2'$, we know, by (7.46), that

$$P(T_1/e) < P(T_2/e) \quad \text{iff} \quad r_1 > r_2.$$

Thus, we have, e.g., the following result for supersessance$_2$:

(7.47)   If α and n are great, $b_1' = b_2' < b$, and $c_1' = c_2'$, then T_1 supersedes$_2$ T_2 in explaining a weak generalization g, given e, just in case $r_1 > r_2$.

So far, we have compared the explanatory power of two theories in L. Let us next take Case (b), in which T_1 is in L and T_2 in $L(\lambda)$. Special interest is attached to the cases in which T_2 is an observational subtheory of T_1. We here consider only the special case in which T_1 does not have non-tautological consequences in $L(\lambda)$ so that T_2 has an empty set of axioms, i.e., T_2 is tautologous in $L(\lambda)$. In other words, we are here interested in the conditions for

(7.48) $syst_i(e, g) < syst_i(T \& e, g), \quad i = 1, 2, 3,$

to hold. (Here $T = T_1$.) When (7.48) holds, the theoretical term 'P' in T can be said to be desirable for inductive explanation in the strong sense that it does some explanatory work not established by any observational subtheory of T.

For $i = 1, 3$, (7.48) holds if and only if

(7.49) $P(g/e) < P(g/e \& T)$.

The same is true for measures $syst_1'$ and $syst_3'$. A number of conditions for (7.49) have been established in Chapters 3, 4, and 5, and summarized in (6.49)–(6.54).

For $i = 2$, (7.48) holds if and only if

(7.50) $P(e)\,[1 - P(g/e)] > P(e \& T)\,[1 - P(g/e \& T)]$.

Again, the same is true for $syst_2'$. Here, by formulas (2.8') and (2.9),

(7.51) $$P(e) = \frac{\sum\limits_{i=0}^{K-c} \binom{K-c}{i} \dfrac{(\alpha + c + i - 1)!}{(n + c + i - 1)!} \prod\limits_{j=1}^{c} (n_j!)}{\sum\limits_{i=0}^{K} \binom{K}{i} \dfrac{(\alpha + i - 1)!}{(i - 1)!}}$$

and

(7.52) $$P(e \& T) = \frac{\sum\limits_{i=0}^{2K-r-c'} \binom{2K - r - c'}{i} \dfrac{(\alpha + c' + i - 1)!}{(n + c' + i - 1)!} \prod\limits_{j=1}^{c'} (n_j'!)}{\sum\limits_{i=0}^{2K} \binom{2K}{i} \dfrac{(\alpha + i - 1)!}{(i - 1)!}}$$

Here $n_j, j=1, \ldots, c$, are the numbers of individuals in the cells of $L(\lambda)$ exemplified in e, and $n'_j, j=1, \ldots, c'$, are the corresponding numbers for cells of L. Notice that, since $x! \geqslant y! \, z!$ when $x = y + z$ and $x \geqslant 1$, we always have

$$\prod_{j=1}^{c'} (n'_j!) \leqslant \prod_{j=1}^{c} (n_j!).$$

If T is an explicit definition of 'P' in terms of λ, then $c' = c$ and $r = K$, so that $P(e \,\&\, T) < P(e)$. Hence, we have

(7.53)   If T is an explicit definition of 'P' in terms of λ, and g is a weak or strong generalization of $L(\lambda)$, then
$$syst_i(e, g) = syst_i(T \,\&\, e, g), \quad i = 1, 3$$
$$syst'_i(e, g) = syst_i(T \,\&\, e, g), \quad i = 1, 3$$
$$syst_2(e, g) < syst_2(T \,\&\, e, g)$$
$$syst'_2(e, g) < syst'_2(T \,\&\, e, g).$$

It is interesting and somewhat unexpected that in the case of $syst_2$ and $syst'_2$ even explicitly definable theoretical concepts do explanatory work. It should be remembered, however, that theories consisting merely of explicit definitions do not inductively explain observational generalizations in the positive relevance sense.

No systematic study is made here of the conditions for (7.50). Since we usually have $P(e \,\&\, T) < P(e)$, (7.50) will hold in most cases in which (7.49) is true, as well as in some others.

Above, we considered the explanatory powers of two rivalling theories with respect to a generalization g of $L(\lambda)$. As another kind of situation, we may mention the case of two competing generalizations g_1 and g_2 of $L(\lambda)$, both compatible with the evidence e. We give here only one simple result:

(7.54)   If $b_1 = b_2$ and $b'_1 < b'_2$, then
$$syst_i(e, g_1) = syst_i(e, g_2)$$
and
$$syst_i(T \,\&\, e, g_1) > syst_i(T \,\&\, e, g_2),$$
for $i = 1, 2, 3$.

In (7.54), T may be a theory in L which does not have non-tautologous consequences in $L(\lambda)$.

4.2. We proceed to study of our measures of systematic power (7.11)–(7.13) which are global with respect to the explanandum. Accordingly, let $\mathbf{D}$ be the set of constituents of $L(\lambda)$. Our discussion is restricted to comparison of the values of $syst_i(e, \mathbf{D})$ and $syst_i(T \& e, \mathbf{D})$, $i = 1, 2, 3$. The principal aim is that of showing that the latter can be greater than the former, even if theory T does not have observational consequences.

According to (7.14)–(7.16), we have

(7.55) $syst_1(e, \mathbf{D}) < syst_1(T \& e, \mathbf{D})$
  iff $\sum_i P(C_i/e) \log P(C_i/e) <$

$$< \sum_i P(C_i/e \& T) \log P(C_i/e \& T)$$

(7.56) $syst_2(e, \mathbf{D}) < syst_2(T \& e, \mathbf{D})$
  iff $P(e)[1 - \sum_i P(C_i/e)^2] >$

$$> P(e \& T)[1 - \sum_i P(C_i/e \& T)^2]$$

(7.57) $syst_3(e, \mathbf{D}) < syst_3(T \& e, \mathbf{D})$
  iff $\sum_i P(C_i/e)^2 < \sum_i P(C_i/e \& T)^2$.

Here $P(e)$ and $P(e \& T)$ are given in formulae (7.49) and (7.50). Note that $syst_3(e, \mathbf{D}) < syst_3(T \& e, \mathbf{D})$ implies $syst_2(e, \mathbf{D}) < syst_2(T \& e, \mathbf{D})$ if $P(e) \geqslant P(e \& T)$.

If $n \to \infty$ and $\alpha \neq \infty$, then the probability of one of the constituents of $L(\lambda)$, relative to both e and to $e \& T$, approaches one. Hence,

$$syst_i(e, \mathbf{D}) \approx syst_i(T \& e, \mathbf{D}) \approx 1,$$

for $i = 1, 2, 3$.

For explicit definitions we have, corresponding to (7.53),

(7.58) If T is an explicit definition of 'P' in terms of λ, then
    $syst_i(e, \mathbf{D}) = syst_i(T \& e, \mathbf{D})$,
    for $i = 1, 3$, but
    $syst_2(e, \mathbf{D}) < syst_2(T \& e, \mathbf{D})$.

However, for piecewise definitions we can have $syst_i(e, \mathbf{D}) < syst_i(T \& e, \mathbf{D})$ also for $i = 1, 3$. This is immediately observable from Example 5.2, e.g., when $n = \alpha$.

Now $syst_1(e, \mathbf{D})$ and $syst_3(e, \mathbf{D})$ have their minimum values, for any e and $\mathbf{D}$, when all the probabilities $P(C_i/e)$ are equal (for constituents C_i compatible with e). This will be the case if $n=\alpha$. Similarly, $syst_1(T \& e, \mathbf{D})$ and $syst_3(T \& e, \mathbf{D})$ have their minimum values if $n=\alpha$ and $d_i=0$, for all C_i compatible with e. In these cases, we have

$$\sum_i P(C_i/e) \log P(C_i/e) = -(K - c)$$

$$\sum_i P(C_i/e \& T) \log P(C_i/e \& T) = -(2K - r - c')$$

$$\sum_i P(C_i/e)^2 = \frac{1}{2^{K-c}}$$

$$\sum_i P(C_i/e \& T)^2 = \frac{1}{2^{2K-r-c'}}.$$

Consequently, the following result is obtained by (7.55) and (7.56).

(7.59)   If $n = \alpha$, then
$$syst_i(e, \mathbf{D}) < syst_i(T \& e, \mathbf{D})$$
holds, for $i=1, 2, 3$, if $K - r - c_0 < 0$.

(That this holds for $i=2$ is trivial from (7.56), (7.51), and (7.52).) Here the condition $K-r-c_0$ requires that T has non-tautological observational consequences.

We can also see that if T rules out some of the constituents C_i of $L(\lambda)$ which are compatible with e then usually $syst_i(e, \mathbf{D}) < syst_i(T \& e, \mathbf{D})$, for $i=1, 2, 3$.

We conclude by giving some simple examples in which T does not have non-tautological observational consequences.

Example 7.1. Let $\lambda = \{O_1, O_2\}$ and $\mu = \{P\}$, and suppose that the Ct-predicates Ct_1, Ct_3, and Ct_4 are exemplified in e (cf. Example 3.1). Then there are two constituents of $L(\lambda)$ compatible with e, viz. C_1, according to which Ct_1, Ct_3, and Ct_4 are instantiated in the universe, and C_2, according to which all the Ct-predicates are instantiated. Suppose further that T is the theory

$$T = (x) [(P(x) \supset O_1(x)) \& (P(x) \supset O_2(x))],$$

and that the individuals of e in Ct_1 split into both Ct_1 & P and Ct_1 & $\sim P$. It can then be proved that

$$syst_i(e, \mathbf{D}) = syst_i(T \& e, \mathbf{D}), \quad \text{if} \quad n = \alpha$$
$$syst_i(e, \mathbf{D}) > syst_i(T \& e, \mathbf{D}), \quad \text{if} \quad n \neq \alpha,$$

for $i = 1, 3$, but

$$syst_2(e, \mathbf{D}) < syst_2(T \& e, \mathbf{D}).$$

Examples in which $syst_i(e, \mathbf{D}) < syst_i(T \& e, \mathbf{D})$ can hold also for $i = 1, 3$, can easily be found, e.g., if $d_i > 0$ at least for one constituent C_j. A situation of this kind is illustrated, e.g., by Example 6.3. Another example will be mentioned here.

Example 7.2. Let $\lambda = \{O_1, O_2\}$ and $\mu = \{P\}$. Suppose that the Ct-predicates Ct_1 and Ct_4 are exemplified in e. There are then four constituents, $C_1, ..., C_4$, of $L(\lambda)$ compatible with e. Here Ct_1 and Ct_4 are instantiated according to C_1, Ct_1, Ct_2, and Ct_4 are instantiated according to C_2, Ct_1, Ct_3, and Ct_4 are instantiated according to C_3, and all Ct-predicates are instantiated according to C_4 (cf. Example 9.6 below). Let T be the theory

$$T = (x)(P(x) \supset O_2(x)),$$

and suppose that the individuals of e in Ct_1 split into both Ct_1 & P and Ct_1 & $\sim P$. Suppose further that $n = \alpha$. Then

$$P(C_i/e) = \tfrac{1}{4}, \quad \text{for} \quad i = 1, ..., 4$$
$$P(C_i/e \& T) = \tfrac{1}{8}, \quad \text{for} \quad i = 1, 2$$
$$P(C_i/e \& T) = \tfrac{3}{8}, \quad \text{for} \quad i = 3, 4.$$

Therefore,

$$\sum_{i=1}^{4} P(C_i/e) \log P(C_i/e) = -2$$

$$\sum_{i=1}^{4} P(C_i/e \& T) \log P(C_i/e \& T) \simeq -1.8$$

$$\sum_{i=1}^{4} P(C_i/e)^2 = \tfrac{1}{4}$$

$$\sum_{i=1}^{4} P(C_i/e \& T)^2 = \tfrac{5}{16}.$$

Then we have, by (7.55) and (7.57),

$$syst_i(e, \mathbf{D}) < syst_i(T \& e, \mathbf{D}),$$

for $i = 1, 3$.

Let n_1 and n_2 be the numbers of individuals of e in Ct_1 and in Ct_4, respectively, and let n_1', n_2', and n_3' be the corresponding number of individuals in $Ct_1 \& P$, $Ct_1 \& \sim P$, and $Ct_4 \& \sim P$. Then, by (7.51) and (7.52),

$$P(e) = 4An_1! \, n_2!$$

and

$$P(e \& T) = 8Bn_1'! n_2'! n_3'!,$$

where

$$\frac{1}{A} = \sum_{i=0}^{4} \binom{4}{i} \frac{(\alpha + i - 1)!}{(i - 1)!}$$

and

$$\frac{1}{B} = \sum_{i=0}^{8} \binom{8}{i} \frac{(\alpha + i - 1)!}{(i - 1)!}.$$

Hence, by (7.46),

$$syst_2(e, \mathbf{D}) < syst_2(T \& e, \mathbf{D})$$

$$\text{iff} \quad 6\binom{n_1}{n_1'} A > 11\, B.$$

Since here $n_1 \neq n_1'$ and $n_1 \geqslant 2$, $\binom{n_1}{n_1'} \geqslant 2$. Since also $A > B$, we have

$$syst_2(e, \mathbf{D}) < syst_2(T \& e, \mathbf{D}).$$

Our next (and last) example shows that evidence e in $L(\lambda \cup \{P\})$, where 'P' is an evidential theoretical predicate, can have more systematic power with respect to $\mathbf{D}$ than e in $L(\lambda)$, even if we do not have any theory in $L(\lambda \cup \{P\})$ at our disposal. This example can be taken to indicate that the reinterpretation of evidence e in the light of a new predicate 'P' can increase the amount of information that e carries concerning $\mathbf{D}$.

Example 7.3. Let λ, μ, e, C_1, and C_2 be as in Example 6.1. If $n = \alpha$, then

$$P(C_1/e) = P(C_2/e) = \tfrac{1}{2}$$

in $L(\lambda)$, but

$$P(C_1/e) = \frac{1}{2^{8-c'}}, \qquad P(C_2/e) = 1 - \frac{1}{2^{8-c'}}$$

in $L(\lambda \cup \{P\})$. Since here $8-c' \geqslant 2$, $syst_i(e, \mathbf{D})$ in $L(\lambda \cup \{P\})$ is greater than $syst_i(e, \mathbf{D})$ in $L(\lambda)$, for $i=1, 3$.

NOTES

[1] It should be emphasized that we discuss here the theoretical inductive explanation of generalizations, rather than the probabilistic explanation of particular facts. In the context of the latter type of inductive explanation, several philosophers have recently argued against Hempel (1962) that it is not high probability, but inductive or statistical relevance which is central for these explanations. See Jeffrey (1969), Rosenkrantz (1969), Greeno (1970, 1971), Salmon (1971a, b).

[2] Hintikka (1968b) discusses the difference between measures of explanatory power which are, in our sense, local with respect to the explanandum and global with respect to the explanans. Greeno (1970) has emphasized the need of measures which are global with respect to the explanandum.

[3] This is no more than a rough outline of the deductive explanation of generalizations by means of theories. For a detailed account, see Tuomela (1972) and (1973).

[4] In Hesse (1968) an explanation is constructed, in which Kepler's second law and Galilei's law of falling bodies are *deductively* explained by means of Newtonian mechanics *and* certain correspondence rules. However, were these correspondence rules at least partially unknown, we would have a typical inductive explanation in our sense. See also our comments on the particular example of inductive explanation given in Chapter 11.4.

[5] As our $syst_2(T \& e, g)$ assumes its minimum when $\sim(T \& e) \vdash g$, it is also based upon an explanation-intuition which is incompatible with the transmitted entropy view.

CHAPTER 8

CORROBORATION AND THEORETICAL CONCEPTS

1. THEORETICAL AND OBSERVATIONAL SUPPORT

One of the central problems in the philosophy of science is the problem of accounting for the *support* that theories and generalizations receive or may receive. This support may come from various sources, which have two main types. One is empirical information obtained by means of observation and experimentation, and the other is theoretical information obtained from other theories and generalizations belonging to the same nomological network. We call these two types of support *observational* and *theoretical support*, respectively.

Recent literature contains a number of approaches to the analysis and measurement of the support of theories and generalizations. These approaches are classifiable as follows.[1] In *non-probabilistic* approaches, the qualitative concept of support received by a theory or a generalization is defined in terms of its deductive consequences (cf. Popper, 1959, p. 266; Stegmüller, 1971, pp. 31–34) or its instances (cf. Hempel, 1945).[2] In the various *probabilistic* approaches, the qualitative and quantitative concepts of support are defined in terms of probabilities. These approaches can be divided into two: those which are *probabilistic in a narrow sense* and those which are *probabilistic in a wide sense*. The former identify the measure of support with posterior probability, while the latter use measures which are only functions of probabilities. Keynes and Reichenbach can be mentioned as representatives of the narrow probabilistic view, and Kemeny and Oppenheim, Finch, Rescher, Levi, and Hintikka of the wide probabilistic view.[3] At least since (1962a), Carnap has also advocated the wide view. The narrow probabilistic approach has been the main target of Popper's 'anti-inductivistic' attack, but his own measures of 'corroboration' do not belong to non-probabilistic (in our sense) but to the wide probabilistic approach.[4]

Even if many philosophers have discussed both observational and theoretical support (cf., for example, Braithwaite, 1953; Hempel, 1966, and Lakatos, 1968), previous technical accounts, including those mentioned above, deal only with observational support. In addition to the fact that many of the above theoreticians have endorsed a narrow form of empiricism, this may be partially attributable to the following two technical aspects of the situation. It has been incorrectly thought that theories and generalizations cannot have a non-zero or high probability. (This is true of both Carnap and Popper; cf. Chapter 11.3.) Secondly, so far no systems of inductive logic have existed that are capable of dealing with theoretical information expressible only by means of theoretical concepts. Our approach in this book avoids these two problems. As can be expected, we emphasize the importance of theoretical support in the evaluation of theories and generalizations. In terms of the above classification, our approach is probabilistic in the wide sense. As the expression 'degree of confirmation' has often been associated with the narrow probabilistic view, we here prefer the phrases 'degree of corroboration' and 'degree of support'.

Before we proceed to specific measures of corroboration, some distinctions need to be discussed. As in earlier chapters, we shall be dealing with a generalization g in $L(\lambda)$, observational evidence e in $L(\lambda)$, and a theory T in $L(\lambda \cup \mu)$. If now $(T \& e) \vdash g$ then we have a case of *hypothetico-deductive* inference. (T is assumed to be a more or less conjectural theory.) Given some evidence e corroborating g, we may ask, for instance, whether and to what extent e corroborates T (either directly or indirectly *via* g). Another type of situation is that in which not $(T \& e) \vdash g$ but rather $(T \& e)Ig$. We are then dealing with *hypothetico-inductive* inference. Naturally, the same questions concerning corroboration can be asked here as were asked in the hypothetico-deductive case. However, we are not primarily concerned below with the corroboration given to T by g and e (see, however, Chapter 11.4), but rather with the degree of corroboration of the empirical theory or generalization g due to empirical evidence e alone, as against the degree of corroboration due to e *and* the *theoretical* support given by T.

The notion of corroboration (depending upon how it is explicated) has many interesting and problematic connections with notions of explanation, acceptance, and verisimilitude (or degree of closeness to truth). Except for a few comments in the text and in Chapter 11.4, we have not here undertaken a systematic study of these connections.

2. Measures of Corroboration Based on Positive Inductive Relevance

2.1.   Sir Karl Popper has repeatedly emphasized that science does not aim at highly probable theories, but at theories with high content, simplicity, testability, and corroborability. These notions are claimed to go together, and to be opposite to high probability. Let us cite Popper himself:

> We want *simple* hypotheses – hypotheses of a high *content*, a high degree of *testability*. These are also the highly *corroborable* hypotheses, for the degree of corroboration of a hypothesis depends mainly upon the severity of its tests, and thus upon its testability. Now we know that testability is the same as high (absolute) logical *improbability*, or low (absolute) logical *probability*.
>
> But if two hypotheses, h_1 and h_2, are comparable with respect to their content, and thus with respect to their (absolute) logical probability, then the following holds: let the (absolute) logical probability of h_1 be smaller than that of h_2. Then, whatever the evidence e, the relative logical probability of h_1 given e can never exceed that of h_2 given e. Thus *the better testable and better corroborable hypothesis can never obtain a higher probability, on the given evidence, than the less testable one.* But this entails that *degree of corroboration cannot be the same as probability.*
>
> This is the crucial result. My later remarks in the text merely draw the conclusion from it: if you value high probability, you must say very little – or better still, nothing at all: tautologies will always retain the highest probability. (Popper, 1959, p. 270.)

Furthermore, for Popper content, logical strength, and explanatory power "amount to one and the same thing" (see Popper, 1963, p. 217, and, 1972, pp. 15–17). He also identifies explanatory power with the degree of corroboration (Popper, 1963, p. 391).

One may severely criticize Popper's lumping together of all of the following notions: logical strength, content, simplicity, improbability, testability, corroborability, explanatory power and degree of corroboration. Popper might try to defend himself by saying that these notions are only extensionally equivalent in the sense that their measures vary together. However, as we shall see, even this is not true. We cannot here enter upon a detailed examination of Popper's methodology of science, but have to restrict ourselves to some remarks on his concept of degree of corroboration.

What does Popper mean by '*degree of corroboration*', except that he considers it to be dependent on simplicity, testability, and improbability? Briefly, for him it is usually a synonym for 'degree of severity of the tests which a theory has passed'; sometimes merely 'degree to which a statement x is supported by a statement y' can be taken as its synonym (cf.

Popper, 1959, p. 391). Popper has devised measures for the degree of corroboration of a theory h, given certain evidence e, where e must be understood to be "a report on the severest tests we have been able to design" (cf. Popper, 1959, p. 418).[5] As is emphasized by Popper (1959, p. 410), these measures are normed versions of the expression

$$(8.1) \qquad P(e/h) - P(e),$$

which represents, *inter alia*, the information carried by h concerning e (our $cont(h///e)$ defined by (6.9); for other interpretations of (8.1), see Hintikka, 1968b).

Popper thinks that the 'severity' of the test (or evidence) e in (8.1) is at least partly reflected in the fact that the more improbable e is, the greater (8.1) is, and that (8.1) becomes the greater the likelier this improbable e is in the light of h. Let us now write '$corr(h, e)$' for the difference (8.1). Then,

$$(8.2) \qquad corr(h, e) = \frac{P(e)\,P(h/e)}{P(h)} - P(e) = \frac{P(e)}{P(h)}(P(h/e) - P(h)).$$

From (8.2) we see that $corr(h, e)$ is *directly* proportional to $P(e)$. Thus, contrary to Popper's intuition, the more probable test e is, the greater $corr(h, e)$ is. Similarly, Popper's normalization of (8.1), viz.

$$(8.3) \qquad \frac{P(e/h) - P(e)}{P(e/h) + P(e)},$$

reduces to

$$\frac{P(h/e) - P(h)}{P(h/e) + P(h)}.$$

This expression does not contain $P(e)$ at all, and thus it fails to reflect the severity of the test e in probabilistic terms.

From (8.2), we also see that $corr(h, e)$ is inversely related to the prior probability of h. Thus, we see that, with $P(e)$ and $P(h/e)$ kept fixed, the initial content of h – and hence its prior improbability as well as all the properties (such as simplicity and logical strength) proportional to the prior improbability of h – directly varies with (8.1), and hence with the degree of corroboration of h. However, $cont(h_1) \geqslant cont(h_2)$ does not imply $corr(h_1, e) \geqslant corr(h_2, e)$. Moreover, the converse claim does not hold either.

The basic reason for this is that even if $P(h_1) \leqslant P(h_2)$, we may have $P(h_1/e) \leqslant P(h_2/e)$ or $P(h_1/e) > P(h_2/e)$, for suitable evidence e. It also follows that Popper's quoted claim concerning the posterior probabilities of h_1 and h_2 on e is false.

As we have seen above, on the one hand Popper identifies content and explanatory power, and on the other hand explanatory power and degree of corroboration. However, our remarks on the lack of direct connection between content and degree of corroboration refutes as least one of Popper's identifications.

Several authors have made the remark in opposition to Popper that it is possible to have high initial content but low content relative to evidence (i.e., high posterior probability) at the same time (see, e.g., Bar-Hillel, 1955).[6] For Michalos (1971), p. 39, this is 'extremely *ad hoc*'. He raises the question: If high initial content is desirable, why is a high relative content not desirable? The answer can be seen immediately from (8.2). It shows that high posterior probability is directly relevant to $corr(h, e)$. In fact, from (8.2) we can see that, for fixed e, a hypothesis which has smaller initial content and larger relative content than all of its competitors will receive the highest degree of corroboration. Within Hintikka's system of inductive logic, the constituent C_c is exactly of this type, whenever the evidence e is sufficiently great.

This shows that the Popperian attack against high probabilities should not be uncritically directed to posterior probabilities. On the other hand, Popper's conclusion that the degree of corroboration is not the same as probability does in general hold good. This is one thing we have found important and acceptable in Popper's work. In this work, we have emphasized the methodological importance of the notion of positive relevance *vis-à-vis* high probability. Consequently, the attempt to base measures of corroboration upon positive relevance or *increase* in probability seems natural to us. This is effected by (8.1), which has the property that $corr(h, e) > 0$ if and only if e is positively relevant to h.

2.2.   Earlier in this book, we compared the probabilities $P(g/e)$ and $P(g/T \& e)$, that is, investigated the difference

(8.4) $P(g/T \& e) - P(g/e)$,

where g is an empirical law in $L(\lambda)$, T is a theory (or theoretical background knowledge) in $L(\lambda \cup \mu)$, and e the total available empirical evidence. Apart from the addition of the evidence e, (8.4) is of the form (8.1). In addition to its connection with explanatory power ((8.4) varies with the explanatory power of T with respect to g, given e, as measured by $syst_3'(T \& e, g)$), (8.4) thus explicates the degree of corroboration that g gives to T in the evidential situation e. A summary of our results of cases in which (8.4) is positive is given by formulas (6.49)–(6.54).

It should be noted that, as a measure of corroboration, (8.4) takes into account only the observational support that theory T receives. Moreover, (8.4) measures the degree of corroboration of T irrespective of whether T entails g or not. If we have a case of hypothetico-deductive inference, so that $(T \& e) \vdash g$, then (8.4) reduces to

$$1 - P(g/e),$$

which equals the content of g relative to e. In other words, if T entails g, given e, then T receives from g corroboration of an amount which equals the content of g relative ot e. However, in this case measure (8.4) trivial-izes in the sense that it cannot reflect any differences between theories (e.g., their content) which entail g. However, in a hypothetico-inductive case where $(T \& e)Ig$, so that $P(g/e \& T) < 1$, (8.4) is not trivial in this sense. Our results in earlier chapters have shown that, in this case, the degree of corroboration of T on g relative to e or (8.4) can be positive, even for theories that lack non-tautological deductive theorems in $L(\lambda)$. This demonstrates a striking difference between deductivistic approaches to support, and approaches that use measures of corroboration based upon positive relevance. In the latter, the support that a theory receives is characterized in terms of its 'inductive observational consequences', viz., by the set $\{g \in \mathbf{D} \mid P(g/T \& e) > P(g/e)\}$, where $\mathbf{D}$ is e.g. the set of consti-tuents of $L(\lambda)$.

Let us now consider the problem of measurement of the observational *and* theoretical support of empirical laws. We thus ask whether the degree of corroboration of an empirical generalization g (in $L(\lambda)$) increases when it is viewed in the light of some background information T (in $L(\lambda \cup \mu)$). Thus, instead of having the evidence e in (8.1) we now have $e \& T$. Accord-ingly, let us put

$$corr_1(g, e, T) = P(e \& T, g) - P(e \& T).$$

This measure is in fact our $cont(g///e \text{ \& } T)$, i.e. (6.9), and it explicates the degree of corroboration of g on the conjunction $T \text{ \& } e$.

We now obtain

(8.5) $corr_1(g, e, T) > corr(g, e)$ iff

$$\frac{P(e \text{ \& } T)\, P(g/e \text{ \& } T)}{P(g)} - P(e \text{ \& } T) > \frac{P(e)\, P(g/e)}{P(g)} - P(e).$$

The probabilities on the left hand side of the inequality are computed in the language L (by a measure P_1, say), and the probabilities on the right hand side in $L(\lambda)$ (by P_0, say). In general, the measures P_0 and P_1 yield different values (e.g. $P_1(g) \neq P_0(g)$), as we know from earlier chapters.

Now we have as an equivalent to (8.5)

(8.6) $corr_1(g, e, T) > corr(g, e)$ iff

$$P_1(e \text{ \& } T)\left[\frac{P_1(g/e \text{ \& } T)}{P_1(g)} - 1\right] > P_0(e)\left[\frac{P_0(g/e)}{P_0(g)} - 1\right].$$

In (8.6) the formulae for $P_1(e \text{ \& } T)$ and $P_0(e)$ are given in Chapter 7 (p. 111); $P_1(g/e \text{ \& } T)$ and $P_0(g/e)$ are obtained from (3.3) and (3.2). The prior probability $P_0(g)$ is obtained from (3.4), and $P_1(g)$ is obtained from (3.3) by putting $r=0$, $b'=2b$, $c'=0$, $n=0$.

We shall not here investigate (8.6) in any detail. For our present purposes, it may suffice to note that the right hand side of (8.6) can hold at least in some cases. To see this, we note that

(8.7) $corr_1(g, e, T) > 0$ iff
$\qquad P_1(e \text{ \& } T/g) - P_1(e \text{ \& } T) > 0$ iff
$\qquad P_1(g/e \text{ \& } T) - P_1(g) > 0.$

Consider now the last expression. It is easily discernible that in general $P_1(g) < P_0(g)$. In addition, in some cases we may well have $P_0(g, e) = P_0(g)$ (e.g. for a $g = C_w, w \neq c$) and at the same time $P_1(g/e \text{ \& } T) > P_0(g/e)$, for some T and e (see Chapter 4 and also Example 5.4). Hence, in such cases $corr_1(g, e, T) > 0$.

Next consider the analogous

(8.8) $corr(g, e) > 0$ iff
$\qquad P_0(g/e) - P_0(g) > 0.$

However, under the present assumptions, $corr(g, e) = 0$, and hence $corr_1(g, e, T) > corr(g, e)$.

The upshot of our discussion so far is that theoretical concepts and theoretical information can increase the corroboration of an empirical or observational generalization (i.e., establish inductive systematization in the sense of corroboration) when corroboration is measured by $corr$ and $corr_1$. This also holds good for Popper's various normed versions of (8.1), as they decrease and increase whenever (8.1) decreases or increases, respectively.

To motivate another measure of corroboration for g, given T and e, recall that the relativized measure (8.4) could be used to measure the amount of corroboration a generalization g gives to a theory T in an evidential situation e:

$$corr_2(T, g, e) = P(g/T \ \& \ e) - P(g/e).$$

In analogy with this, we can consider the problem of corroborating g by T in a given evidential situation e. Let T describe theoretical 'evidence' and/or background assumptions. Then

$$corr_2(g, T, e) = P(T/g \ \& \ e) - P(T/e)$$

explicates the degree of corroboration of g on the theory T, given e. By the measure $corr_2(g, T, e)$ we can compare the effects on the corroboration of g of different theories T in $L = L(\lambda \cup \mu)$. In particular, a theory T may be compared with a tautologous 'theory' T_0 in L. Then, if T_0 is a tautology (in L)

(8.9) $corr_2(g, T, e) > corr_2(g, T_0, e)$ iff
$P_1(T/g \ \& \ e) - P_1(T/e) > P_1(T_0/g \ \& \ e) - P_1(T_0/e).$

Not surprisingly, $corr_2(g, T_0, e)$ equals zero. The question then is whether and when $corr_2(g, T, e)$ is positive. If both g and T are compatible with the evidence e (more exactly: if $P_1(g \ \& \ e) > 0$, $P_1(T \ \& \ e) > 0$), then

(8.10) $corr_2(g, T, e) > corr_2(g, T_0, e)$ iff
$P_1(g/e \ \& \ T) > P_1(g/e).$

It is sufficient for the condition of (8.10) to hold that $r > 0$ and $r + b' \leqslant 2b$, in case P is evidential. At least for large values of α and n, this condition can be replaced simply by $b' < 2b$. This last condition states that T rules out some cells of L which are empty by g. As $P_1(g/e) \leqslant P_0(g, e)$, we are in part led back to previously investigated cases, which show when $P_1(g/e \ \& \ T) > P_0(g/e)$.

Within the present setup, a third way still exists of explicating corroboration. We may ask whether e corroborates g better *in the light of T* than without it. This leads to the following relativized measure

$$corr_3(g, e, T) = P(e/g \ \& \ T) - P(e/T).$$

which explicates just the degree of corroboration of g on e, given T. We now obtain

(8.11) $corr_3(g, e, T) > corr(g, e)$ iff

$$P_1(e/T)\left[\frac{P_1(g/e \ \& \ T)}{P_1(g/T)} - 1\right] > P_0(e)\left[\frac{P_0(g/e)}{P_0(g)} - 1\right].$$

It is sufficient for the condition of (8.11) to hold that $P_1(e/T) \geqslant P_0(e)$ and $P_1(g/e \ \& \ T)P_0(g) > P_0(g/e)P_1(g/T)$.

Here a detailed and general investigation of (8.11) can be bypassed, for we can demonstrate by other means that (8.11) is satisfiable. We can easily see that

(8.12) $corr_3(g, e, T) > 0$ iff

$P(g/e \ \& \ T) - P(g/T) > 0$ iff

$U_2(g, e \ \& \ T) > 0,$

where U_2 is a measure of expected utility defined in Chapter 6. In Chapter 9 we shall see that both in the case weak and strong generalizations g, we can have $U_2(g, e \ \& \ T) > 0$ but $U(g, e) \leqslant 0$ (which equals $corr(g, e) \leqslant 0$) even when T has no observational consequences. This, together with (8.12), shows that theoretical concepts can be logically indispensable for corroborating observational theories by means of some observational evidence and in the light of a background theory.

In the case of $corr_3$, we have just seen that theoretical concepts can be logically indispensable. The same result holds for $corr_1$. In Chapter 9, it will be shown that it is possible to have $U_1(g, e \ \& \ T) > 0$ and $U(g, e) \leqslant 0$

when T has no observational consequences. This implies that in the case of $corr_1$ theoretical concepts can be indispensable, since $U_1(g, e \& T) > 0$ implies $corr_1(g, e, T) > 0$. In the case of $corr_2$, theoretical concepts can be desirable in the sense that $corr_2(g, T, e) > 0$ holds for a theory T in L but $corr_2(g, T', e) = 0$ for all observational subtheories T' of T.

As a general remark on the relationship between corroboration and explanation, let us note the following. When we are discussing the corroboration of T by its inductive consequence g, corroboration and explanation take opposite directions. Thus, in the case of $corr_2$, T inductively explains g, given e, if and only if g corroborates T, given e. However, when an observational generalization g is corroborated by means of a theory T, and the evidence e, explanation and corroboration have, in a sense, the same direction, as T then both explains and corroborates g.

Our discussion above makes it easy to see how to treat more complex methodological situations. Thus, we may ask whether the adoption of a theory T introducing new concepts 'corroboratively' discriminates between two observational generalizations g_1 and g_2 receiving the same degree of corroboration on the basis of some observational evidence e. For instance, by the definitions of $corr_1$, $corr_2$, and $corr_3$, we have

$$corr_1(g_1, e, T) > corr_1(g_2, e, T) \quad \text{iff}$$
$$\frac{P(g_1/e \& T)}{P(g_1)} > \frac{P(g_2/e \& T)}{P(g_2)}$$

$$corr_2(g_1, T, e) > corr_2(g_2, T, e) \quad \text{iff}$$
$$\frac{P(g_1/e \& T)}{P(g_1/e)} > \frac{P(g_2/e \& T)}{P(g_2/e)}$$

$$corr_3(g_1, e, T) > corr_3(g_2, e, T) \quad \text{iff}$$
$$\frac{P(g_1/e \& T)}{P(g_1/T)} > \frac{P(g_2/e \& T)}{P(g_2/T)}.$$

Contrary to Popper, we see again that posterior probabilities (of g_1 and g_2 on e and T) are highly relevant here. The second relevant factor is either the absolute content of g_1 and g_2 ($corr_1$) or their contents relative to e ($corr_2$) or to T ($corr_3$).[7]

To summarize the measures of corroboration introduced in this section, let us recall that

$corr(g, e) = $ the degree of corroboration of g on e

$corr_1(g, e, T) = $ the degree of corroboration of g on e & T

$corr_2(g, e, T) = $ the degree of corroboration of g on T, given e

$corr_3(g, e, T) = $ the degree of corroboration of g on e, given T.

We have examined the following differences:

$corr_1(g, e, T) - corr(g, e)$

$corr_3(g, e, T) - corr(g, e)$

$corr_2(g, e, T) - corr_2(g, e, T_0)$.

The first of these differences compares the corroborative effect of e & T and e, i.e., it gives the increase in corroboration due to T, given e. The second one compares the effect of e in the light of T with the plain effect of e. Thus, it gives another explication of the increase in corroboration due to T, given e. The third one compares the corroborative effect of T with a null theory T_0. It is easy to define other measures of incremental corroboration, and to study their properties on the basis of our results.

3. HINTIKKA'S MEASURE OF CORROBORATION

3.1.   In the preceding section, we studied measures of corroboration based upon positive inductive relevance. They measure support in terms of transmitted information, and were found to be closely connected with our measures of expected utilities. It is clear that these measures do not exhaust our intuitions in regard to corroboration. For instance, they do not respect content and logical strength to the extent one might desire (cf. our quotation of Popper in Section 2). Moreover, they do not adequately measure the degree of truth or verisimilitude of theories and generalizations. Consequently, other measures of corroboration should also be considered.

Below, we work with an explicate of corroboration originally proposed by Hintikka (1968a). Hintikka's measure takes into account both the *enumerative* and *eliminative* aspects of support, as shown in Hintikka (1968a). Hence, it also measures *excess corroboration* in the sense of Lakatos (1968), as will be seen below in more detail. Hintikka's measure satisfies Popper's most important desiderata for corroboration (but, as is reasonable, not all of them).

The basic measure of corroboration to be investigated is the following. Consider a monadic statement (generalization) g in a language $L(\lambda)$ such that g is compatible with the evidence e and $e \vdash g \equiv C_{j_1} \vee \ldots \vee C_{j_g}$. The posterior degree of corroboration of g on the evidence e is defined as

$$corr_4(g, e) = \min_i \{P(C_i/e) \mid i = j_1, \ldots, j_g\}.$$

If g is incompatible with e, we define $corr_4(g/e) = 0$. If g is a strong generalization consisting of only one constituent, its degree of corroboration equals its posterior probability.

The above measure of corroboration can be viewed as follows from the point of view of rational acceptability. As the index of the acceptability of a (weak) generalization g, we use the posterior probability of the least acceptable (i.e. least supported) strong generalization compatible with g and e. (Cf. our discussion in Chapter 9.)

The measure of corroboration defined above satisfies some of Popper's central requirements for a measure of corroboration (cf. Popper, 1963). For instance, if $e \vdash g_1 \supset g_2$, then

$$P(g_1/e) \leqslant P(g_2/e)$$

but

$$corr_4(g_1/e) \geqslant corr_4(g_2/e).$$

Thus, logical strength and $corr_4$ covary.[8] Moreover, for sufficiently large evidence, the constituent with the highest initial content will receive the highest degree of corroboration (cf. Hintikka and Pietarinen, 1966).

It seems that the measure $corr_4$ can serve as an index of verisimilitude or degree of truth of a general statement. Measures of verisimilitude proportional to the posterior probability of a weak generalization give high values to uninformative statements. In contrast, if we are looking for a description of the world which is both informative and true, our $corr_4$ seems to be a much better index. Posterior probability may work as an index of verisimilitude for constituents only. Even in this case, $corr_4$ gives the highest asymptotic value for the initially most informative constituent. This may be regarded as an asymptotic justification for a methodology urging to strive for theories and generalizations with high degree of corroboration (i.e., $corr_4$).

In the similar way to the above, we define, for $L(\lambda \cup \mu)$, the posterior degree of corroboration of g on the evidence e and a theory T by

$$corr_4(g, e, T) = \min_i \{P(C_i^r/e \;\&\; T) \mid i = j_1, ..., j_g\}$$

where

$$e \;\&\; T \vdash g \equiv C_{j_1}^r \vee ... \vee C_{j_g}^r.^9$$

We proceed to study measures $corr_4(g, e)$ and $corr_4(g, e, T)$ with the application of our earlier results in Chapters 2, 3, and 4. We shall assume here that g is a generalization of $L(\lambda)$ which says that as least certain b cells of $L(\lambda)$ are empty. Then the distributive normal form of g contains *all* the constituents of $L(\lambda)$ which require that these b cells are empty.

By formula (2.9),

$$P(C_i/e) = M \frac{(\alpha + w_i - 1)\,!}{(n + w_i - 1)\,!}$$

where M is a constant dependent upon K, c, α, and n, but not on i. Hence we have

$$\alpha < n : P(C_i/e) > P(C_k/e) \quad \text{iff} \quad w_i < w_k$$
$$\alpha > n : P(C_i/e) > P(C_k/e) \quad \text{iff} \quad w_i > w_k$$
$$\alpha = n : P(C_i/e) = \frac{1}{2^{K-c}}, \quad \text{for all} \quad i = j_1, ..., j_g.$$

For all $i = j_1, ..., j_g$, we have $c \leqslant w_i \leqslant K - b$. Hence (2.9) implies for $\alpha < n$ (with $w = K - b$)

$$(8.13) \quad corr_4(g, e) = \frac{1}{\displaystyle\sum_{i=0}^{K-c} \binom{K-c}{i} \frac{(\alpha + c + i - 1)!\,(n + K - b - 1)!}{(n + c + i - 1)!\,(\alpha + K - b - 1)!}},$$

for $\alpha > n$ (with $w = c$)

$$(8.14) \quad corr_4(g, e) = \frac{1}{\displaystyle\sum_{i=0}^{K-c} \binom{K-c}{i} \frac{(\alpha + c + i - 1)!\,(\alpha + c - 1)!}{(n + c + i - 1)!\,(\alpha + c - 1)!}},$$

and for $\alpha = n$

$$(8.15) \quad corr_4(g, e) = \frac{1}{2^{K-c}}.$$

In the special case $\alpha = 0$, $\lambda \to \infty$, the posterior degree of corroboration becomes

$$(8.16) \quad corr_4(g, e) = \frac{1}{\sum_{i=0}^{K-c} \binom{K-c}{i} \left(\dfrac{K-b}{c+i}\right)^n}.$$

In the general case, we have the following results for large values of n and α. When $\alpha < n$:

$$(8.17) \quad corr_4(g, e) \simeq \frac{1}{\sum_{i=0}^{K-c} \binom{K-c}{i} \dfrac{\alpha^{c+i-1} n^{K-b-1}}{\alpha^{K-b-1} n^{c+i-1}}}$$

$$= \frac{\left(\dfrac{\alpha}{n}\right)^{K-b-c}}{\left(1 + \dfrac{\alpha}{n}\right)^{K-c}}$$

$$\to \begin{cases} 0, & \text{if } K - b > c \\ 1, & \text{if } K - b = c \end{cases} \quad \text{when} \quad n \to \infty,$$

and when $\alpha > n$:

$$(8.18) \quad corr_4(g, e) \simeq \frac{1}{\left(1 + \dfrac{\alpha}{n}\right)^{K-c}}.$$

Similarly, in the case of evidential theoretical concepts we know that $c' \leqslant w_i^r \leqslant 2K - r - b'$ for all constituents C_i^r in the distributive normal form of g in L. Hence, we have the following results: for $\alpha < n$:

$$(8.19) \quad corr_4(g, e, T) =$$

$$= \frac{1}{\sum_{j=0}^{2K-c'-r} \binom{2K-c'-r}{j} \dfrac{(n+2K-r-b'-1)! \, (\alpha+c'+j-1)!}{(\alpha+2K-r-b'-1)! \, (n+c'+j-1)!}}$$

$$\simeq \frac{\left(\dfrac{\alpha}{n}\right)^{2K-b'-r-c'}}{\left(1 + \dfrac{\alpha}{n}\right)^{2K-c'-r}} \quad (n \text{ and } \alpha \text{ great})$$

$$\to \begin{cases} 0, & \text{if} \quad 2K - b' - r > c' \\ 1, & \text{if} \quad 2K - b' - r = c' \end{cases} \quad \text{when } n \to \infty,$$

for $\alpha > n$:

$$(8.20) \quad corr_4(g, e, T) = \frac{\dfrac{(\alpha + c' - 1)!}{(n + c' - 1)!}}{\displaystyle\sum_{i=0}^{2K-c'-r} \binom{2K - c' - r}{j} \dfrac{(\alpha + c' + j - 1)!}{(n + c' + j - 1)!}}$$

$$\simeq \frac{1}{\left(1 + \dfrac{\alpha}{n}\right)^{2K-c'-r}} \quad (\alpha, n \text{ great}),$$

and for $\alpha = n$;

$$(8.21) \quad corr_4(g, e, T) = \frac{1}{2^{2K-c'-r}}.$$

These results show the following, *inter alia*. When $n \to \infty$ and $\alpha \neq \infty$, the degree of corroboration of a generalization g approaches one if and only if g is the 'minimal' constituent C_c with respect to observational evidence e or the 'minimal' constituent with respect to e and the theory T (cf. results (8.17) and (8.19)). The importance of this result will be the subject of later comments. Moreover, when $n \neq \infty$ and $\alpha \to \infty$, both $corr_4(g, e)$ and $corr_4(g, e, T)$ approach zero (cf. results (8.18) and (8.20)). This corresponds to Carnap's λ-system.

We can now compare the degree of corroboration of g on e (in $L(\lambda)$) and on e and T (in L). Let us suppose that α and n are great. Then for $\alpha \geqslant n$:

$$(8.22) \quad corr_4(g, e) < corr_4(g, e, T)$$

$$\text{iff} \quad \frac{1}{\left(1 + \dfrac{\alpha}{n}\right)^{K-c}} < \frac{1}{\left(1 + \dfrac{\alpha}{n}\right)^{2K-c'-r}}$$

$$\text{iff} \quad 2K - c' - r < K - c$$

$$\text{iff} \quad K - r < c' - c$$

$$\text{iff} \quad K - r - c_0 < 0,$$

and for $\alpha < n$:

$$(8.23) \quad corr_4(g, e) < corr_4(g, e, T)$$

$$\text{iff} \quad \frac{\left(\dfrac{\alpha}{n}\right)^{K-b-c}}{\left(1+\dfrac{\alpha}{n}\right)^{K-c}} < \frac{\left(\dfrac{\alpha}{n}\right)^{2K-b'-r-c'}}{\left(1+\dfrac{\alpha}{n}\right)^{2K-c'-r}}$$

$$\text{iff} \quad \left(1+\dfrac{n}{\alpha}\right)^{K-r-c_0} < \left(\dfrac{\alpha}{n}\right)^{b-b'}.$$

This holds with $b \geqslant b'$ only if $K-r-c_0 < 0$, and with $b < b'$ only if $K-r-c_0 < b'-b$. These conditions hold only if T has observational consequences. Therefore, for large α and n, a theory T can increase the degree of corroboration of g only if it is observationally strong in the sense that it rules out some cells of $L(\lambda)$. This shows that the measure $corr_4$ is closer to the deductivistic approaches to support than are measures based on positive relevance.

3.2. It is now interesting to ask whether it is possible that for a generalization g,

$$corr_4(g, e) \to 0, \text{ when } n \to \infty$$
$$corr_4(g, e, T) \to 1, \text{ when } n \to \infty.$$

By our results (8.17) and (8.19), this happens if and only if we have

$$K - b > c$$
$$2K - r - b' = c'.$$

In other words, when these two conditions hold, theory T changes the asymptotical degree of corroboration of g from minimum to maximum. (Cf. our criterion CA4 in Chapter 1.3.)

Let us next consider an example which illustrates this asymptotical situation and also shows that $corr_4(g, e, T)$ can exceed $corr_4(g, e)$ preasymptotically. Let g be a generalization

$$g = (x)\,(O_1(x) \supset O_2(x)),$$

in $L(\lambda)$ for which $K=4$. Assume that $c=2$. As $b=1$, we have $K\text{-}b\text{-}c>0$ (as required in the above criterion). The following diagram illustrates g and the evidential situation:

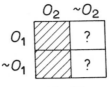

Fig. 8.1.

Let T now be $(x)(\sim O_1(x) \supset P(x))$ & $(x)(P(x) \supset O_2(x))$ in $L(\lambda \cup \{P\})$. (Notice that T has observational consequences as $T \vdash (x)(\sim O_1(x) \supset O_2(x))$.) Now, assuming $c'=3$, we have $K'=2K=8$, $r=4$, and $b'=1$. This is illustrated by the following diagram:

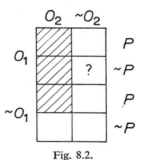

Fig. 8.2.

As $2K-r-b'-c'=0$ and $K-b-c>0$, it follows that $corr_1(g, e) \to 0$, when $n \to \infty$, but that $corr_4(g, e, T) \to 1$, when $n \to \infty$. This change in the asymptotic value of the degree of corroboration of g is due to the fact that T eliminates the constituents of L which claim the subcells (in L) of the cell $\sim O_1$ & $\sim O_2$ of $L(\lambda)$ to be instantiated.

Let us next examine the preasymptotic cases by means of the above formulas for corroboration. In the case $\alpha < n$, (8.13) implies for the present situation

$$corr_4(g, e) = \cfrac{1}{\cfrac{(n+2)!}{(\alpha+2)!}\left[\cfrac{(\alpha+1)!}{(n+1)!} + \cfrac{2(\alpha+2)!}{(n+2)!} + \cfrac{(\alpha+3)!}{(n+3)!}\right]}$$

$$= \cfrac{1}{\cfrac{n+2}{\alpha+2} + 2 + \cfrac{\alpha+3}{n+3}}$$

Correspondingly, (8.19) gives

$$corr_4(g, e, T) = \frac{1}{\frac{(n+2)!}{(\alpha+2)!}\left[\frac{(\alpha+3)!}{(n+3)!} + \frac{(\alpha+2)!}{(n+2)!}\right]} = \frac{1}{1 + \frac{\alpha+3}{n+3}}.$$

Hence, $corr_4(g, e, T) > corr_4(g, e)$ when $\alpha < n$.

For the case $\alpha > n$, we obtain from (8.14) and (8.20)

$$corr_4(g, e) = \frac{1}{1 + 2\frac{\alpha+2}{n+2} + \frac{(\alpha+3)(\alpha+2)}{(n+3)(n+2)}}$$

$$corr_4(g, e, T) = \frac{1}{1 + \frac{\alpha+3}{n+3}}.$$

Hence $corr_4(g, e, T) > corr_4(g, e)$ also when $\alpha > n$. The same result holds also for $\alpha = n$, as then

$$corr_4(g, e) = \tfrac{1}{4}$$
$$corr_4(g, e, T) = \tfrac{1}{2}.$$

It should be noted that in examples of the present kind, T must have observational consequences (cf. above). It follows that we can always find a theory T' in $L(\lambda)$ which has the same effect on the corroboration of g as has T. (In our example $T' = (x)(\sim O_1(x) \supset O_2(x))$.)

A general comment can now be made on the relation between corroboration and elimination. As we have often noted earlier, within Hintikka's inductive logic, generalizations and theories are capable of having maximal asymptotic posterior probabilities, i.e., our condition CA2 of Chapter 1.3 is satisfied. This comes about because one of the constituents (i.e., C_c) in the distributive normal form of the generalization or theory concerned receives a maximal probability value asymptotically. Constituent C_c says that in our universe exactly those kinds of individuals exist which are exemplified in the asymptotical evidence e. This evidence e has eliminated (falsified) all constituents C_w with $w < c$. All constituents C_w ($w > c$) in the distributive normal form in question receive posterior probabilities approaching zero. Hence, if a generalization g is to receive a high degree of corroboration ($corr_4$) in large universes with evidence growing without limit, we must have this: our theoretical background information T has to *eliminate* (falsify) all the constituents C_w $(w > c)$ from the distributive normal form of our generalization g. Thus elimination, both that due to e and that due to T, is a precondition of corroboration in a very strong sense.

Let us view the situation in some more detail. Assume that c^* (the number of kinds of individuals exemplified so far) grows towards c. Then normally some of the constituents in g will be falsified. Let C_c be the widest constituent in g which is not eliminated by T. Now the degree of corroboration of g becomes large asymptotically, or in the case or large finite evidence, if and only if the new evidence exemplifies some of the cells within $c - c^*$. The role of elimination due to the evidence e is here seen in that the exemplification of any cell within $c - c^*$ eliminates those constituents which require that cell to be empty. Thus, asymptotically g will get a high degree of corroboration (i.e., one) if and only if the empirical evidence *cum* T has ruled out all the other constituents besides C_c. We have here a strong requirement of observational *excess corroboration* (cf. Lakatos, 1968, p. 375 ff.): What is important is the corroboration concerning the *novel* claims of the generalization (here: the exemplification of the $c - c^*$ cells) and not the accumulation of evidence of the old and non-discriminative kind.

It should be emphasized that, as our above results show, frequently the *observational* evidence *alone* does not suffice for high (preasymptotic nor even asymptotic) degrees of corroboration. What is again needed is a background theory that usually introduces new or theoretical concepts. This background theory then often takes on a strong *eliminative* role, as we have just seen. Thus, in corroborating a generalization, we have not only the requirement of observational excess corroboration, but also the requirement of falsification due to theory.

That our measure $corr_4$ strongly emphasizes the eliminative role of the background theory T is also reflected in the following fact. Even if T logically implies g, given e, $corr_1(g, e, T)$ need not obtain its maximum value one, and may even approach zero when n grows without limit. However, if g is a constituent of $L(\lambda)$ and $(T \& e) \vdash g$, then we have $corr_4(g, e, T) = 1$. (As in Chapter 4, we assume here that $c_1 = 0$.) But if g is a weak generalization of $L(\lambda)$ its distributive normal form may contain constituents whose probabilities receive asymptotically the probability zero, although T implies g, given e. Therefore, in order to make $corr_4(g, e, T)$ equal to one T has to eliminate these constituents. This is to be seen also in formula (8.19) which gives the value of $corr_4(g, e, T)$ for large α and n such that $\alpha < n$. It can be written in the form

$$\frac{(\alpha/n)^{2K-c'-2b}(n/\alpha)^{r+b'-2b}}{(1+\alpha/n)^{2K-c'-r}}.$$

Since here $n/\alpha > 1$, we see that $corr_4(g, e, T)$ is proportional to the logical strength of T (as measured by the number r of cells of L eliminated by T). What is more, $corr_4(g, e, T)$ is proportional to $r+b'-2b$. This expression gives the number of cells of L which are empty by T but need not be empty by g. Accordingly, this number expresses the *eliminative force* of T with respect to those constituents in the normal form of g which receive asymptotically zero probabilities.

3.3.   In the context considered here, the inductive relation I discussed in Chapter 1.2 may be interpreted as follows (for $j=1, 3, 4$):

(a) $(T \& e \& f)\, Ig$ iff $corr_j(g, e \& f \& T) > corr_j(g, f)$

(b) $(e \& f)\, I\, g$ iff $corr_j(g, e \& f) > corr_j(g, f)$.

In addition to our previous discussion we have here the symbol 'f' to denote suitable observational background evidence. More specifically, f is assumed to state that m individuals have been observed, and have been found satisfy c kinds of individuals. Here e states that $n-m$ additional individuals of the same kind as in f have been observed.

If theory T is to establish inductive systematization (in the sense of Chapter 1.2), then it must be the case that $(T \& e \& f)\, Ig$ (i.e., (a) is true) and not $(e \& f)\, Ig$, (i.e., (b) is false). As we have seen for $corr_j$, $j=1, 3$, this is indeed possible. For $corr_4$ we may have in our earlier example (a) and not (b), so that the example theory T can establish inductive systematization between observational statements. Nevertheless, the theoretical concept P in T is not logically indispensable for corroboration as its subtheory T' (see above) has the same observational corroborative power. For examples of corroboratively indispensable theoretical concepts, see Chapter 9.

Next we have a comment on the corroborative power (in the sense of $corr_4$) of two competing theories with respect to a generalization g, given a certain evidence e. In Chapter 7.4. we discussed a relation of explanatory supersessance S. If we now replace probabilities by degrees of corroboration, we obtain a relation '$S'(T_1, T_2, g, e)$' which reads 'the corroborative force of T_1 with respect to g, given e, is greater than that of T_2'. For the situation discussed in Chapter 7.4, we then get the following results for large values of α and n:

(8.24)   When $\alpha < n$, $S'(T_1, T_2, g, e)$ iff
$$(1 + n/\alpha)^{c'_1 - c'_2 + r_1 - r_2} > (\alpha/n)^{b'_1 - b'_2}$$

(8.25)   When $\alpha \geqslant n$, $S'(T_1, T_2, g, e)$ iff
$$c'_1 - c'_2 + r_1 - r_2 > 0.$$

These results take into account the logical strength of T_1 and T_2 (in terms of r_1 and r_2), the eliminative force of T_1 and T_2 (cf. above), and the variety of evidence (in terms of c'_1 and c'_2). In particular, if $c'_1 = c'_2$ and $b'_1 = b'_2$, then the theory which excludes more cells of L gives g a higher degree of corroboration, for all values of α and n. And if $c'_1 = c'_2$ and $r_1 = r_2$, the theory with a greater eliminative force will give g a higher degree of corroboration, for sufficiently large values of n.

Finally, assume that g_1 and g_2 are two competing generalizations in $L(\lambda)$, both of them compatible with the evidence e, and containing the constituent C_c in their distributive normal forms. If T is a theory in L which is compatible with g_1 and g_2, we have the following results:

(8.26)   If $n \leqslant \alpha$,   then $corr_4(g_1, e, T) = corr_4(g_2, e, T)$

(8.27)   When $n > \alpha$,
$$corr_4(g_1, e, T) > corr_4(g_2, e, T)   \text{iff}   b'_1 > b'_2.$$

When $b_1 = b_2 = b$, this shows that, for sufficiently large evidence, T corroborates g_1 more than g_2, given e, if and only if $r + b'_1 - 2b > r + b'_2 - 2b$, that is, if and only if T has more eliminative force with respect to g_1 than to g_2.

NOTES

[1] Some philosophers prefer to work with a comparative notion of support (see, e.g., Hacking, 1965a, pp. 32–34), and some have claimed that this is the only viable possibility (Kuhn, 1962, pp. 144–146). However, we prefer the quantitative approach for as long at least as most of the important factors in evaluating the support of theories can be incorporated in our measures.

[2] Stegmüller has recently given a definition of *deductive support* which attempts to explicate some of Popper's basic ideas (cf. Stegmüller, 1971, pp. 31–35). This definition is based on the idea that theories can be supported by severe observational test statements which are deducible from it. Apart from the condition of the severity of tests, this definition is essentially the same as the so-called prediction-criterion of confirmation (cf. Hempel, 1945 and Scheffler, 1963). Since this notion of support is characteristic of hypothetico-deductive inference, we leave our comments on it until Chapter 11.

[3] A convenient summary of various proposed measures of support that are probabilistic in the wide sense can be found in Kyburg (1970). See also Hintikka (1968b).

[4] Note that non-probabilistic measures of the degree of support are not included in the above classification. L. J. Cohen has attempted to show that no reasonable measure of support can be a function of probabilities (cf. Cohen, 1971). His proof that there is no satisfactory function f such that the degree of support of a hypothesis h on evidence e equals $f(P(e/h), P(h), P(e))$ has been definitively refuted by Kyburg (1972). It can be added that this proof, even if it had been correct, would not have excluded such probabilistic measures, as our $corr_4$, that depend on the logical structure of the hypothesis h. It should also be emphasized that the probabilities occurring as arguments of a support measure can (and should) depend upon extralogical (and not only *a priori*) factors.

[5] In his latest work, Popper emphasizes stronger than earlier the nature of corroboration as a comparative concept which, after all, is not measurable in a strict sense (cf. Popper, 1972, pp. 82–84). In ideal circumstances, he says, it may be an analytic truth that the degree of corroboration of a hypothesis h_1 in the light of the critical discussion d_t at time t is greater than that of a hypothesis h_2. Even if his emphasis is more on verisimilitude than on corroboration, his basic intuition concerning corroboration seems to be essentially the same as in his earlier works. He still uses such locutions, in connection with corroboration, as 'evidence supports a hypothesis' (cf. Popper, 1972, p. 83).

[6] Somewhat surprisingly, one of Popper's own conditions of adequacy for corroboration (viz., (vi) in Popper, 1959, p. 401) states that the degree of corroboration covaries with posterior probability. This is in sharp contrast with what Popper elsewhere states in so many words.

[7] Our measure $corr_2$ corresponds closely to Mackie's (1969) discussion of confirmation. Thus, $corr_2 (g_1, T, e) > corr_2 (g_2, T, e)$ if and only if T confirms g_1 better than g_2, relative to e, in the sense of Mackie's condition C2. Note, however, that Mackie interprets T as observational evidence. Schlesinger (1970) has argued that Mackie's C2 is not applicable in science. His counterexamples rest on the assumptions that, in our terms, $(e \ \& \ g_1) \vdash T$ and $(e \ \& \ g_2) \vdash T$. Within the present context, these assumptions do not hold, so that Schlesinger's criticism against Mackie cannot applied to measure $corr_2$ as employed here.

[8] In contrast with this idea, Popper has recently claimed that "a statement s which is derivable from a well-tested theory T will, *so far it is regarded as part of T*, have the degree of corroboration of T" (cf. Popper, 1972, p. 20). This is, in fact, a typical condition which probabilistic measures of support (in the narrow sense) do satisfy.

[9] We define $corr_4(g, e, T)$ here as the minimum of the probabilities $P(C_1^r/e \ \& \ T)$, where the C_1^r are those constituents of the richer language L that are compatible with e and T, and occur in the distributive normal form of g in L. We could also define the degree of corroboration of g on e and T by

$$corr_5(g, e, T) = \min \{P(C_i/e \ \& \ T) \mid i = j_1, ..., j_g\}$$

where the minimum is taken over those constituents C_i of the observational language $L(\lambda)$ that are compatible with e and T and occur in the normal form of g in $L(\lambda)$. Since each constituent C_i of $L(\lambda)$ is usually equivalent to a disjunction of several constituents of L, given e and T, we have $corr_5(g, e, T) \geqslant corr_4(g, e, T)$. However, there does not seem to be any important qualitative difference between these two measures. Indeed, it can be verified that our major conclusions about $corr_4$ similarly hold for $corr_5$.

CHAPTER 9

THE LOGICAL INDISPENSABILITY OF THEORETICAL CONCEPTS WITHIN INDUCTIVE SYSTEMATIZATION

1. THE THEORETICIAN'S DILEMMA: METHODOLOGICAL INSTRUMENTALISM REFUTED

According to the doctrine known as scientific instrumentalism, theories and theoretical terms are nothing but logical or conceptual instruments for systematizing our observations and empirical regularities. This metaphor characterizes instrumentalism methodologically. From a semantic viewpoint, instrumentalism claims that theories are non-referring symbolic devices and nothing more. As such, they are not true or false, but only more or less effective.[1]

Unlike the early logical positivists, an instrumentalist can accept the use of 'theoretical' terms which cannot be reduced, by explicit definitions, into the language of experience. In his view, if these terms serve to systematize our observations, they have a kind of pragmatic methodological justification. However, unlike the scientific realists, the instrumentalists do not think of these terms as having cognitive significance or semantic meaningfulness. For them, the theoretical terms do not have evidential uses – they do not refer to unobservable entities or to any reality at all.

There is no need to discuss here all the various arguments which scientific realists can give against instrumentalism (cf., for example, Sellars, 1963; Popper, 1963; and Tuomela, 1973). Instead, we here give a refutation of what can be considered as the strongest argument for instrumentalism. At the same time, this refutation gives strong support to scientific realism.

Let T be a theory in $L(\lambda \cup \mu)$, and let O_T be the set of all deductive observational consequences of T, i.e.,

$$O_T = \{h \text{ in } L(\lambda) \mid T \vdash h\}.$$

It is easy to see that T and O_T establish the same deductive systematization with respect to λ. Moreover, Craig and Ramsey have shown that, under very general conditions on the theory T, we can effectively find a theory, without the theoretical terms μ, which axiomatizes the set O_T. In other words, we have an effective method of eliminating the theoretical terms μ from the theory T by replacing T with one of its subtheories in $L(\lambda)$ which establishes the same deductive systematization with respect to λ as T itself.[2] This shows that there are no theories (satisfying the general conditions of Craig's theorem) of which the theoretical terms are logically indispensable for deductive systematization with respect to λ (see Chapter 1.2).

This result may seem to imply that it is always possible to eliminate theoretical terms from theories without resulting incapability to achieve the purposes of scientific theorizing. If so, there does not seem to exist any reason for assumption that the theoretical terms of any theory are something more than mere non-referring symbolic devices. This argument for instrumentalism, which Hempel (1958) introduced in order to refute it, is known as *the theoretician's dilemma*.

According to the theoretician's dilemma, "the terms and principles of any theory are unnecessary" because if they establish "definite connections among observable phenomena" they are replaceable by purely observational laws linking these phenomena (cf. Hempel, 1958, p. 186). However, from the argument presented above, one cannot make the sweeping conclusion that

(9.1)   Theoretical terms are unnecessary for the purposes of scientific theorizing.

Instead, we have only seen that

(9.2)   Theoretical terms are logically dispensable for deductive systematization with respect to λ.

To derive (9.1) from (9.2), we should need the additional premise that the *only* purpose of theorizing (and of using theoretical terms) is the deductive systematization with respect to λ. However, this premise – even if it appeals to some instrumentalists – is false, however. Theoretical terms are in general introduced to account for and to explain the characteristics and behaviour of the objects, observable or not, which the theory concerns. Thus, besides the deductive systematization with respect to λ, they can serve many different purposes, such as deductive systematization with respect to μ, that is, to other theoretical terms. Moreover, theoretical terms can prove to be logically indispensable for these other purposes of theorizing. In particular, it has been claimed that theoretical terms are logically indispensable for inductive systematization with respect to λ (Hempel, 1958) and for deductive explanation of observational laws (Sellars, 1963; Tuomela, 1973). These claims, if they are valid, show that for certain central aims of science the use of theoretical terms can be essential. In these cases, theoretical terms indeed do some work which is not preserved when the theory is replaced by its Craigian or Ramseyan substitute. This is sufficient to refute the argument of the theoretician's dilemma.

In this chapter, we prove that Hempel's thesis of the logical indispensability of theoretical terms for inductive systematization with respect to λ is correct. Hempel's (1958) original argument for this thesis was unsatisfactory, however, as we argued in Chapter 1.1.[3] Our strategy is thus that of refuting the theoretician's dilemma by solving the transitivity dilemma within our framework. This also shows that the adequacy condition CA7 of Chapter 1.3 is satisfied by our approach.

It will be shown that theoretical terms can be logically indispensable for inductive systematization with respect to λ when the inducibility relation I is interpreted as positive inductive relevance or as inductive acceptance by Hilpinen's (1968) rule A_{cont}. In our examples, we use theories which do not have non-tautological deductive consequences in $L(\lambda)$. Theories of this kind can have only tautological subtheories in $L(\lambda)$. Consequently, if they establish inductive systematization with respect to λ, their theoretical terms are logically indispensable for that purpose according to our definition. Our examples also serve to refute the arguments of Bohnert (1968), p. 280, Hooker (1968), pp. 157–158, Stegmüller (1970), pp. 423, 428–429, and Corman (1972), p. 113 to the effect that theories of this nature are 'empirically trivial' and therefore cannot establish inductive systematization. That these arguments are inconclusive and based on a confusion between the deductive and inductive capabilities of theories (such as testability as opposed to confirmability) has already been shown in Niiniluoto (1973b). (See also Niiniluoto and Hintikka, 1973.)

As was noted in Chapter 1.2, the logical indispensability of theoretical terms does not, as such, suffice to guarantee their methodological usefulness. It is very clearly brought out that gains are obtainable by the use of theoretical terms in the case of deductive systematization, when the role of theoretical terms is considered in conjunction with the problems concerning the depth of theories (cf. Hintikka and Tuomela, 1970; Hintikka, 1973, and Tuomela, 1973). In Chapters 6, 7, and 8 above, we studied some gains obtainable by the use of theoretical terms within inductive systematization. As far as these gains – which were measured by expected epistemic utilities, systematic power, and degrees of corroboration – are obtainable by theoretical terms, they can be called *methodologically desirable*. The results obtained in this chapter, together with those of the three preceding chapters, show that theoretical terms can be logically indispensable as well as methodologically desirable for inductive systematization.

Although these results deprive methodological instrumentalism of most of the plausibility which it may initially seem to possess, someone could still claim that theoretical terms are potentially dispensable in science (see Chapter 1.2). This would be a factual claim which to our knowledge no one has tried to prove. It is, indeed, difficult even to conceive what kind of an argument could be given for its support. As long as such arguments are lacking, we can call theoretical terms not only methodologically desirable but also *methodologically indispensable* in science.

2. LOGICAL INDISPENSABILITY AND POSITIVE INDUCTIVE RELEVANCE

According to our definition in Chapter 1.2, a theory T in $L = L(\lambda \cup \mu)$ establishes inductive systematization with respect to λ and to the relation of positive inductive relevance, given the evidence e, if and only if for some statements k and h of $L(\lambda)$

(A'_1) $P(h/k \,\&\, e \,\&\, T) > P(h/e)$

or

(A'_1) $P(h/k \,\&\, e \,\&\, T) > P(h/e \,\&\, T),$

and

(B) $P(h/k \,\&\, e) \leqslant P(h/e)$

(C) not $(k \,\&\, e \,\&\, T) \vdash h.$

In our examples below, k and e will be singular evidence statements in $L(\lambda)$, h will be a generalization in $L(\lambda)$, and T a theory in L which does not have non-tautological observational consequences in $L(\lambda)$. Then, if the conditions (A'_1) or (A'_2), (B), and (C) hold, the theoretical term 'P' occuring in T is logically indispensable for inductive systematization with respect to λ in the sense of positive relevance.

Example 9.1.   Suppose that $L(\lambda)$ contains three primitive predicates O_1, O_2, and O_3, so that $k=3$ and $K=8$. Let e_n denote the evidence statement according to which we have observed n individuals in cell $Ct_1 = O_1 \,\&\, O_2 \,\&\, O_3$. Let C_w be the constituent of $L(\lambda)$ according to which Ct-predicates Ct_1 and $Ct_2 = O_1 \,\&\, O_2 \,\&\, \sim O_3$ are instantiated in the universe. Then, $c=1$ and $w=2$. Let further $e' = Ct_1(a_{n+1})$, where a_{n+1} is an individual which is not in e_n. Then we have in Hintikka's Jerusalem system, by formula (4.2'),

$$P(C_w/e_n) =$$

$$= \frac{1}{2^n + 7 + 21\left(\frac{2}{3}\right)^n + 35\left(\frac{2}{4}\right)^n + 35\left(\frac{2}{5}\right)^n + 21\left(\frac{2}{6}\right)^n + 7\left(\frac{2}{7}\right)^n + \left(\frac{2}{8}\right)^n}$$

and

$$P(C_w/e_n \,\&\, e') = P(C_w/e_{n+1}).$$

Let T be the theory

$$T = (x)\,[(P(x) \supset O_1(x)) \,\&\, (P(x) \supset O_2(x)) \,\&\, (P(x) \supset O_3(x))]$$

in $L = L(\lambda \cup \{P\})$. Then, $2K = 16$ and $r = 7$. Supposing that 'P' is an evidential theoretical predicate and that $c' = 2$ (i.e., that the individuals in e_n exemplify both $O_1 \,\&\, O_2 \,\&\, O_3 \,\&\, P$ and $O_1 \,\&\, O_2 \,\&\, O_3 \,\&\, \sim P$), we have, $d=0$, $q=3$, and $2K - c' - r = 7$. Hence, by (4.3'),

$$P(C_w/e_n \,\&\, T) =$$

$$= \frac{1}{3^n + 7 + 21\left(\frac{3}{4}\right)^n + 35\left(\frac{3}{5}\right)^n + 35\left(\frac{3}{6}\right)^n + 21\left(\frac{3}{7}\right)^n + 7\left(\frac{3}{8}\right)^n + \left(\frac{3}{9}\right)^n}$$

and

$$P(C_w/e_n \,\&\, e' \,\&\, T) = P(C_w/e_{n+1} \,\&\, T).$$

Letting $n=3$, we have

$$P(C_w/e_3 \ \& \ e') > P(C_w/e_3)$$
$$P(C_w/e_3 \ \& \ e' \ \& \ T) < P(C_w/e_3 \ \& \ T).$$

Hence,

$$P(\sim C_w/e_3 \ \& \ e') < P(\sim C_w/e_3)$$
$$P(\sim C_w/e_3 \ \& \ e' \ \& \ T) > P(\sim C_w/e_3 \ \& \ T),$$

which shows that the conditions (A$'_2$) and (B) are true (for $h = \sim C_w$, $e = e_3$, and $k = e'$).

This example corresponds closely to the Hempelian theory in Chapter 1.1. Notice that, given the evidence e_3, $\sim C_w$ is a weak generalization which contains all constituents of $L(\lambda)$ that are compatible with e_3 except C_w and that C_w would imply the falsity of the generalization $(x) (O_1(x) \ \& \ O_2(x) \supset O_3(x))$.

Example 9.2.   Let $\lambda = \{O_1, O_2\}$, and let e_n denote the evidence that n individuals have been observed to exemplify Ct-predicates $\sim O_1 \ \& \ O_2$ and $\sim O_1 \ \& \ \sim O_2$. Let C_w be the constituent of $L(\lambda)$ according to which the cell $O_1 \ \& \ \sim O_2$ is empty, and let further $e' = \sim O_1(a_{n+1}) \ \& \ O_2(a_{n+1})$ and

$$T = (x) \ [(P(x) \supset O_1(x)) \ \& \ (P(x) \supset O_2(x))]$$

(cf. Example 4.1). If $n = \alpha - 1$, and if P is evidential, then

$$P(C_w/e_n \ \& \ e') = P(C_w/e_n)$$
$$P(C_w/e_n \ \& \ e' \ \& \ T) > P(C_w/e_n \ \& \ T).$$

Thus, the conditions (A$'_2$) and (B) are true. Here C_w implies the generalization $(x) (O_1(x) \supset O_2 (x))$.

Example 9.3.   Let $L(\lambda)$, L, e, C_1, C_2, and T be as in Example 6.4. Suppose that $\alpha = 1$. Then, if $26 < n < 41$,

$$P(C_2/e) < P(C_2) \quad \text{and} \quad P(C_2/e \ \& \ T) \geqslant P(C_2),$$

so that the conditions (A$'_1$) and (B) are true. If $26 < n < 92$, then

$$P(C_2/e) < P(C_2) \quad \text{and} \quad P(C_2/e \ \& \ T) \geqslant P(C_2/T),$$

so that the conditions (A$'_2$) and (B) are true. In this example, C_2 is the constituent of $L(\lambda)$ which allows for all kinds of individuals in the universe. Similar results, with different constraints for n, can be proved in the event that P is a non-evidential theoretical term.

Example 9.4.   Let $\lambda = \{O_1, O_2\}$, and let C_w say that cell $\sim O_1$ & $\sim O_2$ is empty. Suppose that we have observed n individuals in cell O_1 & O_2 (evidence e), and let T be the following piecewise definition of P:

$$T = (x)\,(P(x) \equiv O_1(x)) \vee (x)\,(P(x) \equiv O_1(x) \;\&\; O_2(x))$$

(cf. Example 5.2). Then

$$P\,(C_w/e \;\&\; T) > P\,(C_w/e)$$

holds for all values of n and α. However, since we have here $w > c$, $P\,(C_w/e) = P\,(C_w)$ will hold for suitably chosen values of n and α. Then, for these values of n and α, we also have

$$P\,(C_w/e \;\&\; T) > P\,(C_w),$$

so that the conditions (A_1') and (B) are true.

These examples suffice to show that theoretical terms may be logically indispensable, in the positive relevance sense, with respect to both weak (Example 9.1) and strong (Examples 9.2–9.4) generalizations C_w, with $w > c$, of $L(\lambda)$. (The constituent C_c cannot play the role of h in the condition (B), however, since $P\,(C_c/k \;\&\; e) > P\,(C_c/e)$ holds in Hintikka's system, whenever C_c says that there are only that kind of individuals which are already exemplified in e and k is compatible with C_c.) As theories establishing this inductive systematization we used theories of the Hempelian type (Examples 9.1–9.3) and piecewise definitions (Example 9.4), which contained evidential or non-evidential theoretical predicates.

If we say, in the manner of Carnap (1962a), that k *confirms* h relative to e if and only if

$$P\,(h/k \;\&\; e) > P\,(h/e),$$

i.e., if and only if k is positively relevant to h relative to e, then our results in this section can be formulated as follows: It is possible that k does not confirm h relative to e but, at the same time, 'k & T' confirms h relative to e and, further, k confirms h relative to e and T, where T is a theory in $L(\lambda \cup \mu)$ which has only tautological subtheories in $L(\lambda)$, and k, h, and e are in $L(\lambda)$.

In view of the connection we have established earlier, the results of this section can also be interpreted in terms of expected epistemic utilities (see Chapter 6.2.3), of inductive explanation (cf. Chapter 7) and of degrees of corroboration (see Chapter 8.2). Thus, theoretical terms of Examples 2, 3, and 4 can be said to be logically indispensable for inductive explanation and for corroboration.

3. Logical Indispensability and Rules of Acceptance

3.1.   Among the philosophers and statisticians who have discussed the nature of inductive inference there is a sharp contrast between the *behaviouralists* (including Carnap and Jeffrey) and the *cognitivists* (including Levi and Kyburg).[4] The behaviouralists hold that inductive inference leads only to an assignment of probabilities to hypotheses and that these probabilities can then be utilized by the practical decision-maker. The cognitivists defend the indispensability of inductive rules which allow for the acceptance or rejection of hypotheses, and conceive of scientific knowledge as a body of hypotheses which are rationally accepted as (presumably) true.

An interesting unification of these two approaches is provided by the application of decision-theory to the pragmatics of science, viz. the analysis of scientific inference in terms of the maximization of the epistemic utilities of the scientist (see the works of Hempel, Levi, Hintikka, Hilpinen, and Pietarinen, referred to in Chapter 6). The potentialities in the conceptualization of various aspects of inductive systematization in information-theoretic and decision-theoretic terms should also be evident from Chapter 6. It is important to note that our technical treatment of inductive systematization is, broadly speaking, impartial in respect to the controversy between behaviouralism and cognitivism. In allowing the explication of inductive inference to vary with different kinds of inductive situations, we hope to have captured something of philosophical interest both to the behaviouralists and to the cognitivists.

Thus, our information-theoretic point of departure in Chapter 6 has lead us, in an especially natural way, to stress the importance of the notion of positive inductive relevance in inductive systematization in general and in inductive explanation in particular. This notion is intelligible on the sole premise that we can assign inductive probabilities to the statements of the languages we are considering. Consequently, all our results involving the notion of positive relevance should be perfectly acceptable and interesting to the behaviouralistically minded theorists of induction. At the same time, it is consistent with our emphasis on the methodological significance of the notion of positive relevance, to appreciate the cognitivist analysis of many central aspects of scientific method, such as the acceptance and rejection of scientific theories.

In the preceding section, it was shown that (evidential as well as non-evidential) theoretical terms can be logically indispensable for inductive systematization with respect to λ and to the relation of positive relevance. A cognitivist may consider this result as fairly weak, as positive relevance is a weak explicate of the relation of inducibility. In this section however, we show that the same conclusion can also hold good for inductive acceptance rules.

Inductive acceptance rules have been proposed by many philosophers, *inter alia*, by Hempel, Kyburg, Levi, Hintikka, Hilpinen, and Lehrer. We consider only two of these rules here: rule AG_n, proposed in Hintikka and Hilpinen (1966) and elaborated further in Hilpinen (1968), and rule A_{cont}, proposed in Hilpinen (1968). Not only are these rules readily applicable within our framework, but they also possess further merits and advantages over other rules. AG_n is a rule of acceptance for generalizations within Hintikka's system of inductive logic, while A_{cont} is a decision-theoretic rule of acceptance lacking some undesirable features of Levi's (1967a) rule (see the criticism of Levi in Hilpinen, 1968, pp. 99–104, and Hilpinen, 1972).

Rules AG_n and A_{cont} take into account both observational evidence and the posterior probability of a hypothesis or its informational content. Below, we extend these rules so that they also take into account theoretical background information (in L) in addition to observational evidence (in $L(\lambda)$). When these are the only discriminating factors, rules AG_n and A_{cont} can be viewed as characterizing the acceptance of hypotheses as (presumably) true.

3.2. *Rule AG_n applied to $L(\lambda)$ and L.* A basic feature of Hintikka's system of inductive logic is that when the amount of evidence grows without limit the probabilities of all constituents but one, namely C_c, approach zero. In other words,

(9.3)
$$\lim_{n \to \infty} P(C_c/e) = 1$$
$$\lim_{n \to \infty} P(C_w/e) = 0, \quad \text{when} \quad w > c,$$

where, as usual, n is the size of the sample e and c is the number of Ct-predicates exemplified in e. Let $0 < \varepsilon < 0.5$, and define

$$\varepsilon' = \frac{1}{1 - \varepsilon} - 1,$$

so that

$$1 - \varepsilon = \frac{1}{1 + \varepsilon'}.$$

Then we have in $L(\lambda)$, by (4.2),

$$P(C_c/e) > 1 - \varepsilon$$

if and only if

$$\varepsilon' > \sum_{i=1}^{K-c} \binom{K-c}{i} \frac{(n+c-1)!\,(\alpha+c+i-1)!}{(\alpha+c-1)!\,(n+c+i-1)!}.$$

If we now define

$$n_0 = \text{the largest integer } n \text{ for which}$$

(9.4)
$$\varepsilon' \leqslant \sum_{i=1}^{K-c} \binom{K-c}{i} \frac{(n+c-1)!\,(\alpha+c+i-1)!}{(\alpha+c-1)!\,(n+c+i-1)!},$$

then

(9.5) $P(C_c/e) > 1 - \varepsilon \quad \text{iff} \quad n > n_0.$

Since $1 - \varepsilon > 0.5$, the condition $n > n_0$ implies that C_c is the most probable of all constituents of $L(\lambda)$, given e. In this case, rule AG_n tells us to accept C_c and all its deductive consequences. If $Ac(g/e)$ is allowed to mean that generalization g of $L(\lambda)$ is acceptable on e, rule AG_n can be stated as follows:

(AG$_n$) $Ac(g/e)$ iff $n > n_0$ and C_c occurs in the distributive normal form of g.

Thus, when $n > n_0$, the only constituent which is acceptable on e by AG$_n$ is C_c. According to AG$_n$, it is necessary and sufficient for the acceptance of a generalization g is that the posterior probability of g given the evidence e is high and that the size of the evidence e is sufficiently large.

The extension of rule AG$_n$ for the acceptance of generalizations of $L(\lambda)$ within the richer language $L = L(\lambda \cup \{P\})$ is straightforward. In correspondence with (9.3), for the probabilities of the constituents C_w of $L(\lambda)$ within L we have

(9.6)
$$\lim_{n \to \infty} P(C_w/e \ \& \ T) = 1, \quad \text{if} \quad w = c$$
$$\lim_{n \to \infty} P(C_w/e \ \& \ T) = 0, \quad \text{if} \quad w > c,$$

for theories T in L which are compatible with C_w (cf. (4.7)). If we now define

$$n_1 = \text{the largest integer } n \text{ for which}$$

(9.7) $$\varepsilon' \leqslant \sum_{i=1}^{2K-r-c'} \binom{2K - r - c'}{i} \frac{(n + c' - 1)! \, (\alpha + c' + i - 1)!}{(\alpha + c' - 1)! \, (n + c' + i - 1)!},$$

then, in correspondence with (9.5),

(9.8) $P(C_c/e \ \& \ T) > 1 - \varepsilon$ iff $n > n_1$.

By virtue of (9.8), we can now define rule AG$'_n$, which extends AG$_n$ to L, as follows:

(AG$'_n$) $Ac(g/e \ \& \ T)$ iff $n > n_1$ and C_c occurs in the distributive normal form of g,

where of course $Ac(g/e \ \& \ T)$ means that g is acceptable on the basis of the evidence e and the theory T.

If we have

(9.9) $n_0 > n_1$,

then, for values of n such that $n_0 > n > n_1$,

$$Ac(C_c/e \ \& \ T), \quad \text{but not} \quad Ac(C_c/e).$$

To show the logical indispensability of theoretical terms for inductive systematization with respect to rules AG_n and AG'_n, we should need an example where (9.9) holds and the theory T does not have deductive consequences in $L(\lambda)$. This is not possible, however:

(9.10)   If the theory T does not have non-tautological deductive consequences in $L(\lambda)$, then $n_0 \leqslant n_1$.

For the proof of (9.10), first note that the right hand sides of (9.4) and (9.7) decrease when n grows, and that $n_0 > n_1$ can hold only if the right hand side of (9.7) is less than that of (9.4), for suitable values of n. If $\alpha = n$ or if α and n are large, then $n_0 > n_1$ can hold only if

$$2K - r - c' < K - c$$

which is equivalent to

$$K - r - c_0 < 0$$

and (since $K - r - c_0 = b' - b$) to

$$b' < b.$$

This last condition can hold only if T has non-tautological deductive consequences in $L(\lambda)$. Similarly, if $c' = c$, then $n_0 > n_1$ can hold only if $b' < b$. When $c' > c$ and $n \neq \alpha$, $n_0 > n_1$ can hold only for such values of n that $n < \alpha$. The condition $n < \alpha$ implies, however, that C_c is the least probable of all constituents of $L(\lambda)$ which are compatible with e and that $P(C_c/e)$ as well as $P(C_c/e \ \& \ T)$ are less than 0.5, and conversely. For these values of n, no generalizations are acceptable by AG_n or by AG'_n.

If we let ε also take values greater than 0.5, (9.10) is not generally true. This implies that we allow for the acceptance of generalizations with probabilities of less than 0.5. In other words, we should violate the so called *coherence principle* for acceptance rules:

$$Ac(h/e) \quad \text{only if} \quad P(h/e) \geqslant P(\sim h/e).$$

This shows that if we accept the coherence principle, as seems reasonable, then theoretical terms (of any theory) cannot be proved to be logically indispensable for inductive systematization with respect to λ by the application of rule AG_n. Thus, our general conclusion about the rule AG_n is essentially negative.

This negative conclusion has another aspect, however. It is possible that T is a theory in L that has no non-tautological deductive consequences in $L(\lambda)$ and, at the same time, $n_0 < n_1$. Then, for values of n such that $n_0 < n < n_1$, we have

$$Ac(C_c/e), \quad \text{but not} \quad Ac(C_c/e \ \& \ T).$$

If g is now a generalization in $L(\lambda)$ of which the distributive normal form contains C_c, then g is acceptable on e by AG_n, but not acceptable on e and T by AG'_n. In this case, theory T may be said to establish inductive systematization with respect to λ *in a negative sense*: T reveals that the generalization g which seems acceptable on e within $L(\lambda)$ is not acceptable in the light of the information embodied in the richer language L and in T.

3.3.   *Rule A_{cont} applied to $L(\lambda)$ and L.* In Chapter 6, we defined the expected utility of accepting a generalization g of $L(\lambda)$ on the basis of evidence e by

(9.11) $E(g/e) = P(g/e) - P(g)$

(see formula (6.44)). We also saw that (9.11) follows if the relevant utilities are measured in terms of the total informational content of g, i.e., $cont(g)$, or of the information g carries on e, i.e., $cont(g//e)$.

To follow Hilpinen (1968), (9.11) may be generalized by introducing an additional index q, $0 \leqslant q \leqslant 1$, in it:

(9.12) $E(g/e) = P(g/e) - qP(g)$.

If the distributive normal form of g contains the constituents $C_{j_1}, C_{j_2}, \ldots, C_{j_g}$, that is, if

$$\vdash_{L(\lambda)} g \equiv C_{j_1} \lor C_{j_2} \lor \ldots \lor C_{j_g},$$

then the expected utility of accepting g on e is, by (9.12),

(9.13) $E(g/e) = \sum_{i=j_1}^{j_g} [P(C_i/e) - qP(C_i)]$.

According to the principle of maximizing expected utility, we should accept the generalization of $L(\lambda)$ which maximizes (9.13). If more than one generalization of this kind exists, we choose the one with the least content relative to e. This principle yields the following acceptance rule:

(A_{cont})   Reject all constituents C_w of $L(\lambda)$ with $P(C_w/e) < qP(C_w)$, and accept the disjunction h^* of all unrejected constituents as strongest on the basis of e. Accept all deductive consequences of h^*, and no other generalizations.

A_{cont} tells us to reject all constituents against which we have strong evidence. If we call unrejected constituents *plausible* constituents, then a generalization g of $L(\lambda)$ is acceptable on e, by A_{cont}, if and only if its distributive normal form contains *all* plausible constituents. More formally, let $\mathbf{D}$ be the set of all constituents of $L(\lambda)$, and let $\mathbf{A}(e)$ be the set of constituents plausible on e, that is,

(9.14) $\mathbf{A}(e) = \{C_w \text{ in } \mathbf{D} \mid P(C_w/e) \geqslant qP(C_w)\}$.

If we now define h^* by

(9.15) $h^* = \bigvee_{C_w \in \mathbf{A}(e)} C_w$,

then rule A_{cont} can be stated in the following form:

(A_{cont}) $Ac(g/e)$ iff $h^* \vdash g$
             iff   the distributive normal form of g contains all C_w in $\mathbf{A}(e)$.

Index q has a natural interpretation as an index of the *degree of boldness*. The greater q is, the more difficult is it to reject constituents, and the greater is the risk of error in accepting generalizations. Similarly, the less q is, the more cautious the investigator is. (See Levi, 1967a.)

Rule A_{cont} can be extended to L in two different ways. In correspondence with (9.12), the expected utility of accepting g on e and T, where T is a theory in L, can be defined either by

(9.16) $E_1(g/e \& T) = P(g/e \& T) - qP(g)$,

where $P(g)$ is the probability of g in $L(\lambda)$, or by

(9.17) $E_2(g/e \& T) = P(g/e \& T) - qP(g/T)$

(cf. formulae (6.45) and (6.46)). In correspondence with (9.14) and (9.15), define

(9.18)
$$A_1(e, T) = \{C_w \text{ in } \mathbf{D} \mid P(C_w/e \,\&\, T) \geqslant qP(C_w)\}$$

$$A_2(e, T) = \{C_w \text{ in } \mathbf{D} \mid P(C_w/e \,\&\, T) \geqslant qP(C_w/T)\}$$

and

(9.19)
$$h_1^* = \bigvee_{C_w \in A_1(e,\, T)} C_w$$

$$h_2^* = \bigvee_{C_w \in A_2(e,\, T)} C_w .$$

The acceptance of generalizations g of $L(\lambda)$ on the basis of e and T can then be defined as follows:

(A'_{cont}) $Ac_1(g/e \,\&\, T)$ iff $h_1^* \vdash g$

                        iff   the distributive normal form of g contains all C_w in $A_1(e, T)$

        $Ac_2(g/e \,\&\, T)$ iff $h_2^* \vdash g$

                        iff   the distributive normal form of g contains all C_w in $A_2(e, T)$.

Here $Ac_1(g/e \,\&\, T)$ explicates the acceptance of g on the basis of '$e \,\&\, T$', while $Ac_2(g/e \,\&\, T)$ explicates the acceptance of g on the basis of e relative to the theory T.

Note that when $q=1$, we have, for the measures U, U_1, and U_2 of Chapter 6,

$$C_w \in A(e) \qquad\quad \text{iff} \quad U(C_w/e) \geqslant 0$$

$$C_w \in A_1(e, T) \quad \text{iff} \quad U_1(C_w/e \,\&\, T) \geqslant 0$$

$$C_w \in A_2(e, T) \quad \text{iff} \quad U_2(C_w/e \,\&\, T) \geqslant 0.$$

Rules A_{cont} and A'_{cont} can be generalized, in a straightforward way, to apply to the acceptance of generalizations relative to an evidential situation or background evidence as expressed by a statement b, say, of $L(\lambda)$. Then definitions (9.14) and (9.18) should be replaced by

(9.20)
$$A(e, b) = \{C_w \text{ in } \mathbf{D} \mid P(C_w/e \,\&\, b) \geqslant qP(C_w/b)\}$$

$$A_1(e, T, b) = \{C_w \text{ in } \mathbf{D} \mid P(C_w/e \,\&\, b \,\&\, T) \geqslant qP(C_w/b)\}$$

$$A_2(e, T, b) = \{C_w \text{ in } \mathbf{D} \mid P(C_w/e \,\&\, b \,\&\, T) \geqslant qP(C_w/b \,\&\, T)\}.$$

Rules A_{cont} and A'_{cont} have the following connections to measures $corr_4$ and $corr_5$ of corroboration, respectively (cf. Chapter 8, note 4). If g is acceptable on e, relative to b, then $corr_4(g, e \ \& \ b) \geqslant q \ corr_4(g, b)$. If g is acceptable on e and T, relative to b, then $corr_5(g, e \ \& \ b, T) \geqslant q \ corr_4(g, b)$. If g is acceptable on e, relative to T and b, then $corr_5(g, e \ \& \ b, T) \geqslant q \ corr_5(g, b, T)$.

Rule A_{cont} is asymptotically equivalent to AG_n, for a suitable choice of q (see Hilpinen, 1968, p. 118). However, the flexibility in the choice of the index q gives us the opportunity of showing that theoretical terms, evidential as well as non-evidential, can be logically indispensable for inductive systematization with respect to λ and to A_{cont}. This will be proved in the following examples.

Example 9.5. Let $\lambda = \{O_1, O_2\}$, and let e_n describe a sample of n individuals exemplifying Ct-predicates $Ct_1 = O_1 \ \& \ O_2$, $Ct_3 = \sim O_1 \ \& \ O_2$, and $Ct_4 = \sim O_1 \ \& \ \sim O_2$. Let C_1 be the constituent of $L(\lambda)$ according to which Ct-predicates Ct_1, Ct_3, and Ct_4 are instantiated in the universe, and C_2 be the constituent according to which all Ct-predicates of $L(\lambda)$ are instantiated. Let T be the theory

$$T = (x)[(P(x) \supset O_1(x)) \ \& \ (P(x) \supset O_2(x))],$$

and suppose that P is an evidential theoretical predicate such that the individuals in e_n exemplify both $Ct_1 \ \& \ P$ and $Ct_1 \ \& \ \sim P$. (See Figure 9.1.)

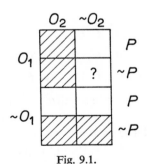

Fig. 9.1.

Thus, we have the same situation as that in Example 6.4 and that in Example 9.3.

If $q=1$ and $26<n<41$, or if $q=0.5$ and $59<n<83$, then

(9.21) $A(e_n) = \{C_1\}$ and $A_1(e_n, T) = \{C_1, C_2\}$.

Similarly, if $q=1$ and $26<n<92$, or if $q=0.5$ and $59<n<197$, then

(9.22) $A(e_n) = \{C_1\}$ and $A_2(e_n, T) = \{C_1, C_2\}$.

When (9.21) holds, we have

$$Ac(C_1/e_n), \text{ but not } Ac_1(C_1/e_n \And T)$$
$$\text{not } Ac(C_2/e_n), \text{ and not } Ac_1(C_2/e_n \And T)$$
$$Ac(C_1 \vee C_2/e_n), \text{ and } Ac_1(C_1 \vee C_2/e_n \And T).$$

The only conclusion to be drawn from these results is that T establishes inductive systematization between e_n and C_1 in a negative sense (cf. above).

This situation is illustrative of the nature of rule A_{cont}. Since, in Hintikka's system, we always have $P(C_c/e)>P(C_c)$, constituent C_c (C_1 in our example) will always belong to the set $A(e)$. Therefore, to find a situation in which

$$\text{not } Ac(C_1/e), \text{ but } Ac_i(C_1/e \And T) \ (i = 1, 2)$$

we need a case where

(9.23) $A(e) = \{C_1, C_2\}$ and $A_i(e, T) = \{C_1\}$

holds instead of (9.21) and (9.22).

To show that such cases can be found, let e_n play the role of our background evidence (i.e., b in (9.20)), and let e_m say that m new individuals are found to exemplify Ct-predicates Ct_1, Ct_3, and Ct_4 (not necessarily all of them). Then

$$P(C_1/e_n) = \frac{1}{1 + \dfrac{\alpha + 3}{n + 3}}$$

$$P(C_2/e_n) = \frac{\alpha + 3}{n + 3} P(C_1/e_n)$$

$$P(C_1/e_n \And e_m) = \frac{1}{1 + \dfrac{\alpha + 3}{n + m + 3}}$$

$$P(C_2/e_n \ \& \ e_m) = \frac{\alpha + 3}{n + m + 3} \ P(C_1/e_n \ \& \ e_m)$$

$$P(C_1/e_n \ \& \ T) = \frac{1}{1 + \dfrac{\alpha + 4}{n + 4}}$$

$$P(C_2/e_n \ \& \ T) = \frac{\alpha + 4}{n + 4} \ P(C_1/e_n \ \& \ T)$$

$$P(C_1/e_n \ \& \ e_m \ \& \ T) = \frac{1}{1 + \dfrac{\alpha + 4}{n + m + 4}}$$

$$P(C_2/e_n \ \& \ e_m \ \& \ T) = \frac{\alpha + 4}{n + m + 4} \ P(C_1/e_n \ \& \ e_m \ \& \ T).$$

By these formulae, we see that

$$\mathbf{A}(e_m, e_n) = \{C_1, C_2\} \quad \text{iff} \quad q \leqslant \frac{\alpha + n + 6}{\alpha + n + m + 6}$$

(9.24) $\mathbf{A}_1(e_m, T, e_n) = \{C_1\}$ iff $q \leqslant \dfrac{(n + m + 4)(\alpha + n + 6)}{(n + 3)(\alpha + n + m + 8)}$ and

$$q > \frac{(\alpha + 4)(\alpha + n + 6)}{(\alpha + 3)(\alpha + n + m + 8)}.$$

These three constraints on q can all be satisfied at the same time if and only if

$$n + m < \alpha.$$

In other words, when $n + m < \alpha$ and the index q satisfies the constraints in (9.24), we have

not $Ac(C_1/e_n \ \& \ e_m)$, but $Ac_1(C_1/e_n \ \& \ e_m \ \& \ T)$.

If g is the generalization

$$g = (x)(O_1(x) \supset O_2(x)),$$

this result shows also that

not $Ac(g/e_n \ \& \ e_m)$, but $Ac_1(g/e_n \ \& \ e_m \ \& \ T)$, since

$$e_n \vdash g \equiv C_1.$$

A corresponding claim does not hold for Ac_2, however, since $A(e_m, e_n) = \{C_1, C_2\}$ implies here that $A_2(e_m, T, e_n) = \{C_1, C_2\}$.

If P is a non-evidential theoretical predicate, then

$$P(C_1/e_n \,\&\, T) = \frac{1}{1 + \dfrac{v_0 (\alpha + 4)}{v_1 (n + 4)}},$$

where

$$v_0 = 2 \frac{n + 3}{\alpha + 3} + n_1 + 1$$

$$v_1 = 2 \frac{n + 4}{\alpha + 4} + n_1 + 1,$$

and n_1 is the number of individuals of e_n exemplifying Ct_1 (cf. formula (3.18)). Further, $P(C_2/e_n \,\&\, T) = 1 - P(C_1/e_n \,\&\, T)$. Again, it can be shown that P is logically indispensable for inductive systematization between e_m and C_1, given e_n, with respect to A_{cont}, with suitable values for q, if and only if $n + m < \alpha$.

Even if this example elegantly illustrates the nature of rule A_{cont}, it has the undesirable feature that the condition $n + m < \alpha$ implies that $P(C_1/e_n \,\&\, e_m)$ and $P(C_1/e_n \,\&\, e_m \,\&\, T)$ are less than 0.5, so that the coherence principle is violated. This is not, however, an inherent feature of A_{cont} (as it was seen to be of AG_n). This is shown by our next example.

Example 9.6. Let $\lambda = \{O_1, O_2\}$ and let e_n and e_m be two samples (with different individuals) exemplifying Ct_1-predicates $Ct = O_1 \,\&\, O_2$ and $Ct_4 = {\sim}O_1 \,\&\, {\sim}O_2$. Let $Ct_2 = O_1 \,\&\, {\sim}O_2$ and $Ct_3 = {\sim}O_1 \,\&\, O_2$, and define constituents C_1–C_4 by

$$C_1 = (Ex)Ct_1(x) \,\&\, (Ex)Ct_4(x) \,\&\, (x)\,(Ct_1(x) \lor Ct_4(x))$$

$$\begin{aligned} C_2 = (Ex)Ct_1(x) \,\&\, (Ex)Ct_2(x) \,\&\, (Ex)Ct_4(x) \,\&\, \\ \&\, (x)\,(Ct_1(x) \lor Ct_2(x) \lor Ct_4(x)) \end{aligned}$$

$$\begin{aligned} C_3 = (Ex)Ct_1(x) \,\&\, (Ex)Ct_3(x) \,\&\, (Ex)Ct_4(x) \,\&\, \\ \&\, (x)\,(Ct_1(x) \lor Ct_3(x) \lor Ct_4(x)) \end{aligned}$$

$$\begin{aligned} C_4 = (Ex)Ct_1(x) \,\&\, (Ex)Ct_2(x) \,\&\, (Ex)Ct_3(x) \,\&\, (Ex)Ct_4(x) \,\&\, \\ \&\, (x)\,(Ct_1(x) \lor Ct_2(x) \lor Ct_3(x) \lor Ct_4(x)). \end{aligned}$$

Here C_1–C_4 are all constituents of $L(\lambda)$ compatible with e_n, and evidence e_n is 'symmetric' with respect to C_2 and C_3. Let T be the theory

$$T = (x)\,(P(x) \supset O_2(x)),$$

where P is evidential, and suppose that the individuals of e_n split into both Ct_1 & P and Ct_1 & $\sim P$. (See Figure 9.2.)

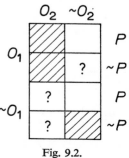

Fig. 9.2.

Since T implies that a 'half' of Ct_2 is empty, one can expect that it favours C_3 to C_2.

We have now

$$P(C_1/e_n) = \cfrac{1}{1 + 2\,\dfrac{\alpha + 2}{n + 2} + \dfrac{(\alpha + 2)\,(\alpha + 3)}{(n + 2)\,(n + 3)}}$$

$$P(C_2/e_n) = P(C_3/e_n) = \frac{\alpha + 2}{n + 2}\,P(C_1/e_n)$$

$$P(C_4/e_n) = \frac{(\alpha + 2)\,(\alpha + 3)}{(n + 3)\,(n + 3)}\,P(C_1/e_n)$$

$$P(C_1/e_n \ \& \ T) =$$

$$= \cfrac{1}{1 + 3\,\dfrac{\alpha + 3}{n + 3} + 3\,\dfrac{(\alpha + 3)\,(\alpha + 4)}{(n + 3)\,(n + 4)} + \dfrac{(\alpha + 3)\,\alpha + 4)\,(\alpha + 5)}{(n + 3)\,(n + 4)\,(n + 5)}}$$

$$P(C_2/e_n \ \& \ T) = \frac{\alpha + 3}{n + 3}\,P(C_1/e_n \ \& \ T)$$

$$P(C_3/e_n \ \& \ T) = \frac{\alpha + 3}{n + 3}\left(2 + \frac{\alpha + 4}{n + 4}\right) P(C_1/e_n \ \& \ T)$$

$$P(C_4/e_n \ \& \ T) = \frac{(\alpha + 3)(\alpha + 4)}{(n + 3)(n + 4)}\left(2 + \frac{\alpha + 5}{n + 5}\right) P(C_1/e_n \ \& \ T).$$

Applying these formulas, we see that

$C_1 \in A(e_m, e_n)$ always

$C_2 \in A(e_m, e_n)$ iff $C_3 \in A(e_m, e_n)$

$$\text{iff} \quad q \leqslant \frac{\dfrac{n + 2}{\alpha + 2} + 2 + \dfrac{\alpha + 3}{n + 3}}{\dfrac{n + m + 2}{\alpha + 2} + 2 + \dfrac{\alpha + 3}{n + m + 3}} = b\,(\text{say})$$

$C_4 \in A(e_m, e_n)$ iff $q \leqslant \dfrac{n + 3}{n + m + 3} b = c\,(\text{say}).$

Here $0 \leqslant c < b$, and $b < 1$ iff $(\alpha+2)(\alpha+3) < (n+3)(n+m+3)$. Consequently,

If $0 \leqslant q \leqslant c$, then $A(e_m, e_n) = \{C_1, C_2, C_3, C_4\}$.
If $c < q \leqslant b$, then $A(e_m, e_n) = \{C_1, C_2, C_3\}$.
If $b < q \leqslant 1$, then $A(e_m, e_n) = \{C_1\}$.

These results clearly illustrate the role of q as an index of the degree of boldness, and also the fact that evidence $e_n \ \& \ e_m$ does not effect a discrimination between C_2 and C_3 in $L(\lambda)$.

If we define

$$d_i = \frac{P(C_i/e_n \ \& \ e_m \ \& \ T)}{P(C_i/e_n)},$$

for $i = 1, 2, 3, 4$, then

$C_i \in A_1(e_m, T, e_n)$ iff $q \leqslant d_i$.

Here

$$d_2 = \frac{(\alpha + 3)(n + 2)}{(\alpha + 2)(n + m + 3)} d_1$$

$$d_3 = \left(2 + \frac{\alpha + 4}{n + m + 4}\right) d_2$$

$$d_4 = \frac{(\alpha + 4)(n + 2)(n + 3)}{(\alpha + 2)(n + m + 3)(n + m + 4)}\left(2 + \frac{\alpha + 5}{n + m + 5}\right) d_1.$$

For fixed n and α, when m grows without limit, d_1 approaches the value $1/P(C_1/e_n)$, while d_2, d_3, and d_4 approach zero. Moreover, we always have $d_2 < d_3$, and, for sufficiently great m,

$$d_4 < d_2 < d_3 < d_1.$$

The values of $d_i = 1, 2, 3, 4$, can now be compared with the values of b and c above. What is important for our purposes is that it is possible to have

$$d_2 < b < d_3.$$

For an example, let $\alpha = n = 2$ and $m = 20$. Then

$$P(C_i/e_2) = \tfrac{1}{4}, \quad \text{for} \quad i = 1, 2, 3, 4$$

$$P(C_1/e_2 \ \& \ e_{20}) = \tfrac{30}{41}$$

$$P(C_2/e_2 \ \& \ e_{20}) = P(C_3/e_2 \ \& \ e_{20}) = \tfrac{5}{41}$$

$$P(C_4/e_2 \ \& \ e_{20}) = \tfrac{1}{41}$$

$$P(C_1/e_2 \ \& \ e_{20} \ \& \ T) = \tfrac{585}{1024}$$

$$P(C_2/e_2 \ \& \ e_{20} \ \& \ T) = \tfrac{117}{1024}$$

$$P(C_3/e_2 \ \& \ e_{20} \ \& \ T) = \tfrac{261}{1024}$$

$$P(C_4/e_2 \ \& \ e_{20} \ \& \ T) = \tfrac{61}{1024}.$$

Hence,

$$b = \tfrac{20}{41}$$

$$c = \tfrac{4}{41}$$

$$d_1 = \tfrac{585}{256}$$

$$d_2 = \tfrac{117}{256}$$

$$d_3 = \tfrac{261}{256}$$

$$d_4 = \tfrac{61}{256}.$$

(See Figure 9.3.)

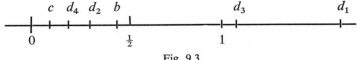

Fig. 9.3.

Consequently, for values of q such that

$$d_2 < q \leqslant b,$$

we have

(9.25)
$$A(e_{20}, e_2) = \{C_1, C_2, C_3\}$$
$$A_1(e_{20}, T, e_2) = \{C_1, C_3\}.$$

Let g be the generalization

$$g = (x)\,(O_1 x \supset O_2 x)$$

of $L(\lambda)$. Since

$$e_2 \vdash g \equiv (C_1 \vee C_3),$$

the result (9.25) implies that

$$\text{not } Ac(g/e_2 \,\&\, e_{20}), \text{ but } Ac_1(g/e_2 \,\&\, e_{20} \,\&\, T),$$

by the rules A_{cont} and A'_{cont}. In other words, the theoretical term P is logically indispensable for inductive systematization between e_{20} and g, given e_2. Moreover, we have

$$P(g/e_2 \,\&\, e_{20} \,\&\, T) = \tfrac{423}{512} \simeq 0.83,$$

so that the coherence principle is not violated.

NOTES

[1] Cf. Nagel's (1961) discussion. As advocates of various formulations of the instrumentalist view, Nagel mentions Peirce, Dewey, Ramsey, Schlick, Ryle, and Toulmin. A number of contemporary physicists could be mentioned as well (cf. Bunge, 1969).
[2] For details and discussions of Craig's and Ramsey's elimination methods, see, for example, Craig (1953, 1956 and 1960), Putnam (1965), Ramsey (1931), Hempel (1958), Bohnert (1961, 1967, and 1968), Scheffler (1963 and 1968), Carnap (1966), Maxwell (1970), Sneed (1971), Stegmüller (1969), Tuomela (1973), Hintikka and Niiniluoto (1973), Niiniluoto (1973a), and Cornman (1972).

[3] Lehrer's (1969) attempt to show that theoretical terms can be logically indispensable for inductive systematization concerns inductive-probabilistic systematization, and thus will not be discussed in this book. That his attempt is also unsatisfactory has been argued in Niiniluoto (1971 and 1973a), and in Tuomela (1971).

[4] See Carnap (1962b and 1963), Jeffrey (1956), Levi (1967a, b), Kyburg (1968), and Hilpinen (1968).

CHAPTER 10

LINGUISTIC VARIANCE IN INDUCTIVE LOGIC

The technical approach to hypothetico-inductive inference which was developed in the preceding chapters of this book is directly and indirectly relevant to many important problems in the philosophy of science. Some of these problems have already been mentioned above. For example, we showed how theoretical concepts can be desirable and even logically indispensable for the purposes of scientific theorizing. In Chapter 9, we indicated how this can be taken to support scientific realism against methodological instrumentalism – and, of course, against those 'descriptive' or 'positivist' views which do not allow of open theoretical concepts assuming any role in science.

Our approach is also relevant to many philosophical problems about the inductive aspects of scientific inference. We have already applied our framework to discussion of the nature of inductive explanation of laws and the corroboration of laws. In this chapter, we consider in more detail the problem of connecting conceptual change with induction, which in Chapter 1 served as a starting point for our approach. In doing this, we contrast our approach with some current views on the desirability of 'linguistic invariance' in inductive logic, and with some current theories of 'probability kinematics'. Furthermore, we have devoted one section to discussion of Goodman's much debated 'new riddle of induction', some aspects of which our results seem to clarify.

1. LINGUISTIC INVARIANCE AND LINGUISTIC VARIANCE

1.1. A serious and largely unsettled problem, which concerns each system of inductive logic, is the extent to which the inductive relation between a hypothesis and evidence should be dependent on the language in which they are stated. Another way of posing this question is to ask what general features of inductive inference should be *linguistically invariant*, and what features can be functions of the choice of language. In particular, we may ask what features, if any, of the choice of language should have an effect on inductive probabilities.

We are not familiar with any single study in which this problem of linguistic invariance as opposed to linguistic variance, in its general form, has been discussed in a completely satisfactory way. Furthermore, many of its philosophical and technical aspects remain to be clarified and worked out. In this section, we make a brief comment on some current views about this problem, and indicate how our approach in this work is related to it. The closely related subject of probability kinematics will be discussed in the next section.

As is well known, in both Carnap's λ-system and Hintikka's λ-α-system, the inductive probability of a singular prediction is computed as a kind of weighed average of an empirical factor and a logical factor (see p. 23 above). The former is determined by observed relative frequences, while the latter is determined by the choice of primitive predicates of the language concerned. This makes these systems sensitive to the choice of primitive predicates and, in particular, to their number. This dependence of inductive probabilities upon the number of primitive predicates of the language has been considered as undesirable, and a counterintuitive feature, by many theorists of induction (e.g., Salmon, 1961a; Nagel, 1963; and Hacking, 1969).

In his article 'Vindication of induction' (1961a), Wesley Salmon proposes the following *criterion of linguistic invariance* as a condition of adequacy for inductive rules:

(CLI₁)   No inductive rule is acceptable if the results it yields are functions of the arbitrary features of the choice of language.

Since, according to Salmon's view, "the choice of a language and consequently the choice of basic predicates is largely arbitrary" (Salmon, 1961a, p. 249), he interprets the criterion of linguistic invariance CLI₁, in effect, in the following way:

(CLI₂)   No inductive rule is acceptable if the results it yields are functions of the choice of primitive predicates.

Salmon shows, further, that the methods of estimation (of limits of relative frequences) based on Carnap's λ-system violate the criterion CLI₂, and concludes that Carnap's inductive logic is inadequate 'for dealing with facts'.

In his *Logical Foundations of Probability* (1950), Carnap had endorsed a requirement of descriptive completeness. This principle asserted that when we apply a language to a given universe, for the purposes of inductive logic, the primitive predicates of this language should be so chosen that they are sufficient to express all the qualitative attributes exhibited by the individuals in this universe (and nothing else). If this principle could be accepted, then the choice of the primitive predicates would not be an arbitrary matter. In other words, we could then claim that Carnap's system does not violate the criterion CLI_1, and that we have no reason to accept the criterion CLI_2.

Carnap also defended his earlier position by pointing out that inductive probabilities in his system depend on the number k of primitive predicates only through the number K of Q-predicates (where $K=2^k$ when the primitive predicates are logically independent), and that for any given universe there in only one adequate value of K (Carnap, 1952, p. 48; see also Nagel, 1963, pp. 795–797). However, Carnap later abandoned the requirement of descriptive completeness, and adopted a condition which makes inductive probabilities dependent only upon those predicates which occur in the hypothesis and the evidence in question, but not upon other families of predicates in the language (see $A11$ in Carnap, 1963, p. 975). That also this solution of the problem of linguistic invariance is unsatisfactory has been argued in Salmon (1967). Moreover, Hacking (1969) has shown how Carnap's systems can be made independent of the choice of primitive predicates within each family, that is, linguistically invariant in the sense of the criterion CLI_2.

But why, after all, should we accept the criterion CLI_2? It surely does not follow from the plausible criterion CLI_1 alone. Indeed, a consideration of the arguments which have been offered for CLI_2 shows that they are not as convincing as is often thought.

For Salmon, the criterion of linguistic invariance has been a part of his programme of 'vindicating induction'. In (1961a), Salmon hoped to be able to show that the only asymptotic inductive rule (for estimating the limits of relative frequences) which satisfies the normalizing conditions and the criterion CLI_1 is the so-called *straight rule*. This rule recommends us to make our estimate of the limit of relative frequency equal to the observed relative frequency. Thus, it relies solely on the empirical factor, and disregards the logical factor. The straight rule is a special case of Carnap's λ-continuum, with $\lambda=0$, and is the only member of this continuum which is linguistically invariant in the sense of CLI_2.

However, criterion CLI_2 does not suffice to vindicate the straight rule. Although it eliminates all the inductive rules which depend on the logical factor, asymptotic rules depending on the sample in a different way from the straight rule can still satisfy CLI_2 (cf. Hacking, 1965b). What is more, relying on Goodman's riddle of 'grue' emeralds, Barker (1961a) has argued that the straight rule does not satisfy criterion CLI_1 either. The obvious 'moral' to be drawn from this result seems to be that Salmon's criterion of linguistic invariance itself is defective (cf. Rudner, 1961). Salmon (1961b) admits that his criterion needs some additional restrictions regarding the type of language to be considered. In (1963a), he tries to escape the Goodman paradox by restricting the use of the straight rule to 'purely ostensive predicates'. This implies that induction may be concerned only with statements expressible by means of observational predicates of a special kind.

Thus we see that Salmon's criterion of linguistic invariance does not have those consequences he initially wanted and thought it had. It does not help him to vindicate any single inductive rule, and to save his own favourite rule, the straight rule, from his own criterion he has to restrict the field of application of inductive rules to the simple cases with 'purely ostensive predicates'. This consequence seems utterly unsatisfactory to all those who try to develop systems of inductive logic that are, at least to some extent, realistic or able to deal with theories and theoretical predicates. (A modest attempt in this direction is made in this book.) At the same time, this conclusion makes the criterion of linguistic invariance seem extremely open to suspicion.

But perhaps we do not really have a choice? Even if this criterion does not have consequences that would satisfy us, Salmon has tried to show that its rejection leads to 'outrageous metaphysics' and further to logical inconsistency. We try to evaluate the force of these two claims, beginning with the former one.

1.2. Salmon's contention that the rejection of the criterion of linguistic invariance leads to outrageous metaphysics is based on his view that "the alternative to adopting such criterion is to regard some particular language as having a privileged status for description and inference" (Salmon, 1961b, p. 260) though the choice of language is, in fact, 'largely arbitrary'. Now it is easy to agree with Salmon that there does not exist any particular language which has such a status. We can also admit that such general rules for the choice of language as Carnap's (1950) requirement of descriptive completeness or Lehrer's (1963) rule of greater completeness are ultimately indefensible. It seems clear that there is no logic, in a strict sense, for the choice of language. In other words, the choice of language is not a matter of logic alone, and cannot be solved within any system of inductive logic. Still, this choice is much less arbitrary than Salmon seems to think. The choice of language is a strongly context-dependent process, where the context may involve such factors as the particular universe which is the object of the study, the particular hypothesis for which we are trying to find evidence, and theoretical knowledge and theoretical presuppositions concerning the constitution of the universe.[1] Thus, normally there are strong pragmatic reasons for tentatively preferring one language to another within a given context (see also Michalos, 1971, pp. 66–69). Though pragmatic in nature, these reasons need not be 'arbitrary' or 'metaphysical'. Consequently, Salmon's step from the criterion CLI_1 to the criterion CLI_2 is not justified.

This point can be elaborated further. It may suffice here to point out the following. The choice of language is an extremely important aspect within the context of scientific discovery. As has frequently been emphasized, progress in science often comes about through conceptual revolutions, that is to say, through the application of new conceptual schemes to reality. It has been plausible argued that conceptual schemes are already in work in the process of perception, and that they already have an important role in determination of what the 'facts' are. This implies that the traditional distinction between language (conceptual system) and facts is simplified and misleading (cf. Hanson, 1958; Kuhn, 1962, and Scheffler, 1967). The criterion of linguistic invariance, when interpreted in a strong sense, seems to express, in an extreme form, this faith in objective facts independent of any conceptual categories. For example, Salmon speaks of 'objective facts' of which the probabilities should not "change with the language in which we choose to talk about them" (Salmon, 1961a, p. 249). Nevertheless, if there are no language-independent facts, at least such that are relevant for science, then the basic motive for adopting a strong criterion of linguistic invariance (as CLI_2 above) vanishes.

It is, of course, desirable that inductive methods should not depend upon arbitrary features of languages. The crucial question, however, is precisely which features of languages are really *arbitrary*.[2] In view of the dependence of facts upon conceptual systems, it seems that the systems of inductive logic have to tolerate a certain amount of conceptual or *linguistic variance*, and, as Rudner (1961, pp. 263–264) puts it, "an obvious desideratum is a theory which will eliminate or minimize unwanted *consequences* of such inevitable variance."

In his reply to the discussion on his paper 'On Vindicating Induction' (1963a), Salmon remarks that the fundamental justification for his criterion of linguistic invariance "is not that it is intuitively plausible and not that it avoids metaphysics," but that it is a consistency requirement (Salmon, 1963b, p. 51). In (1963a), p. 30, he formulates this criterion in the following form:

(CLI$_3$)   Whenever two inductive inferences are made according to the same rule, if the premises of the one differ purely linguistically from the premises of the other, then the conclusion of the one must not contradict the conclusion of the other.

Essentially similar formulations of this criterion can be found in Salmon (1966, p. 102). The principle CLI$_3$ is, of course, ambiguous for as long as we do not specify what is meant by an 'inductive inference made according to a rule' and by a 'purely linguistic difference' between sentences. It is clear that CLI$_3$ allows for many interpretations in this respect.

According to Salmon (1963a, p. 29), two expressions in a given language L have a purely linguistic difference if their equivalence follows from the syntactical and semantical rules of L, that is, when they are L-equivalent. Let us write $\vdash_L e \equiv e'$ in case e and e' are L-equivalent sentences of L. The criterion CLI$_3$ can now be interpreted, for instance, in the following ways.

(CLI$_4$)   If $\vdash_L e \equiv e'$ and a sentence g of L is acceptable on e, then the sentence $\sim g$ must not be acceptable on e'.

(CLI$_5$)   If $\vdash_L e \equiv e'$ and e supports a sentence g of L, then also e' supports g.

(CLI$_6$)   If $\vdash_L e \equiv e'$, then $P(g/e) = P(g/e')$, for each sentence g of L.

All of these criteria are consistency requirements, in an obvious sense.

CLI_4 is natural condition of adequacy for inductive rules of acceptance, CLI_5 corresponds to Hempel's Equivalence Condition for qualitative confirmation (cf. Hempel, 1945), and CLI_6 to Carnap's condition of L-equivalent evidences (cf. Carnap, 1950, p. 285). Moreover, all of these three straightforward interpretations of the criterion CLI_3 are satisfied by inductive probabilities in Carnap's and Hintikka's systems and by suitably defined rules of acceptance and relations of qualitative confirmation within these systems. This assertion needs one qualification, however. If L-equivalence includes equivalences based on definitions or 'semantical rules', so that the primitive predicates of L need not be logically independent, we have to compute all the probabilities in these systems conditional on the truth of the meaning postulates which express, in L itself, the definitory connections between the primitive predicates of L (cf. Carnap, 1952b and 1971b). Then the criteria CLI_4, CLI_5, and CLI_6 are (or can be) satisfied, which shows that Carnap's and Hintikka's systems have reasonable properties of linguistic invariance, even if they do not satisfy the criteria CLI_2. When we apply these systems within one language, no violations of CLI_3 can arise.

Salmon wishes to interpret his criterion CLI_3 in a stronger sense, however. His aim is to show that systems which violate the criterion CLI_2 lead to *interlinguistic contradictions* which are demonstrable in the metalanguage (Salmon, 1963b, p. 52). For that purpose, he defines two expressions in two *different* languages L_1 and L_2 as having a purely linguistic difference if their equivalence is demonstrable in the metalanguage containing the syntax and semantics of both L_1 and L_2 (Salmon, 1963, pp. 29–30; also see Carnap, 1956, pp. 57–58). Moreover, he argues that principles corresponding to CLI_4, CLI_5, and CLI_6, with the notion of intralinguistic L-equivalence replaced by the notion of interlinguistic L-equivalence (i.e., purely linguistic difference between sentences in different languages), should be satisfied by systems of inductive logic.

However, what could these interlinguistic contradictions be like? To gain more insight for this problem, let us consider an example by means of which Michalos (1971), pp. 89–90, has tried to show that Hintikka's system violates Salmon's criterion of linguistic invariance CLI_3 (cf. Niiniluoto, 1973c).

Let L_1 be a monadic language with primitive predicates 'R' and 'B', and let L_2 be a monadic language with primitive predicates 'R', 'U', and

'M'. Let g_1 and g_2 be the generalizations

$$g_1 = (x)\,(R(x) \supset B(x))$$

and

$$g_2 = (x)\,(R(x) \supset U(x)\ \&\ M(x))$$

in L_1 and in L_2, respectively. Suppose that in a sample e of five individuals cells '$R\ \&\ B$', '$\sim R\ \&\ B$', and '$\sim R\ \&\ \sim B$' are exemplified (with respect to L_1), and that in a sample e' of five individuals cells '$R\ \&\ U\ \&\ M$', '$\sim R\ \&\ U\ \&\ \sim M$', '$\sim R\ \&\ \sim U\ \&\ M$', and '$\sim R\ \&\ \sim U\ \&\ \sim M$' are exemplified (with respect to L_2). Assuming that $\alpha = 0$ and $\lambda(w) = w$, we have in Hintikka's system

$$P\,(g_1/e) = .727$$

in L_1, and

$$P\,(g_2/e') = .287$$

in L_2. Let us now interpret the predicates of L_1 and L_2 by the following semantical rules:

(10.1)
'Rx' means that x is Roman
'Bx' means that x is a bachelor
'Ux' means that x is unmarried
'Mx' means that x is male.

Since the expression 'bachelor' is defined as 'unmarried male', it can be assumed that the five individuals in the samples e and e' are, in fact, the same. Moreover, it can be argued that the generalizations g_1 and g_2 differ from each other purely linguistically in Salmon's sense; and, since they have different probabilities on the basis of the same sample of individuals, we seem to have here a genuine case of an interlinguistic contradiction.

There are two problems in this argument, however. The first one is that though the *individuals* in e and e' are the same, the evidence *statements* e and e' are still different. In applying L_2, we observe unmarried and married males and females, while in applying L_1 we observe bachelors and non-bachelors. Hence, the data e and e' are not the same. The second problem, which interests us more here, is that the generalizations g_1 and g_2 are not logically equivalent. Nor does their equivalence follow from the four semantical rules (10.1), given above. They cannot be L-equivalent with respect to either of the languages L_1 and L_2, and to prove their (interlinguistic) equivalence in the metalanguage we should need the additional premise that

(10.2)   The expressions 'bachelor' and 'unmarried male' have always the same extension.

Thus, it is no wonder that the probabilities of g_1 and g_2 are different in L_1 and L_2, since there is nothing in the generalizations themselves (as expressed in L_1 and L_2, respectively), nor in the evidence relative to which their probabilities are computed, that would guarantee their equivalence.

It seems to us that the approach to inductive logic developed in this work gives us a reasonable method for dissolving 'interlinguistic contradictions' of this kind. In the first place, it is important to note that the metalinguistic premise (10.2) and the semantical rules (10.1) have a different status. The latter attach meanings to the linguistic symbols 'R', 'B', 'U', and 'M', while the former asserts that the extensions of the predicates 'bachelor' and 'unmarried male' are identical. Thus, (10.2) has the force of an explicit definition:

(10.3) $(x) (B(x) \equiv U(x) \,\&\, M(x))$.

This definition (let us call it 'h') can be expressed neither in L_1 nor in L_2. For that purpose, we need a language, say L_3, which contains all the predicates 'R', 'B', 'U', and 'M'. Even in L_3 the definition h is not a logical truth, but can be adopted in L_3 as an analytically true *meaning postulate*. Following Carnap's suggestion that all confirmation values be computed as conditional to the truth of the meaning postulates (cf. Carnap, 1952b and 1971b), we have in L_3

$$P(g_1/e \,\&\, h) = P(g_2/e' \,\&\, h) = .287,$$

since, in L_3,

$$h \vdash g_1 \equiv g_2.$$

and

$$h \vdash e \equiv e'.$$

In short, when we take into account, on the level of the object language, the premise which asserts the equivalence of g_1 and g_2, and of e and e', the alleged contradiction disappears.

An interesting feature of this example is that the probability of g_1 in L_3 on the basis of e and h is the same as the probability of g_1 in L_2 on the basis of e'. This is, of course, a direct consequence of the general result (5.1) which we have established for explicit definitions within Hintikka's system of inductive logic.

The basic features of this example can be expressed as follows in a more general way. Let L_1 and L_2 be two monadic languages with the same set of individual constants and with sets $\lambda_1 = \{O_1^1, ..., O_k^1\}$ and $\lambda_2 = \{O_1^2, ..., O_m^2\}$ of logically independent primitive predicates, respectively. Let SR_1 and SR_2 be sets of metalinguistic semantical rules attaching meanings to the predicates in λ_1 and in λ_2, respectively (cf. (10.1)). Let MP be the set of metalinguistic statements which express the analytical connections between the meanings attached to λ_1 and λ_2 by SR_1 and SR_2. Let L_3 be the language with the same individual constants as L_1 and L_2 and with the set $\lambda_1 \cup \lambda_2$ of primitive predicates. Let MP_0 be the set of statements of L_3 which correspond to the metalinguistic statements in MP. This set MP is the set of meaning postulates for the language L_3.

We shall say that the language L_1 is *translatable* into the language L_2 if the predicates in λ_1 are explicitly definable in terms of the predicates in λ_2, by the meaning postulates MP_0. The languages L_1 and L_2 are *inter-translatable* if L_1 is translatable into L_2 and vice versa.[3]

If L_1 is translatable into L_2, then for each sentence g of L_1 there is a *translation* $t(g)$ in L_2, so that g and $t(g)$ are equivalent by the metalinguistic principles MP. If the equivalence of g and $t(g)$ is based solely on such simple postulates in MP which equate the extension of a primitive predicate of L_1 with the extension of a primitive predicate of L_2 (that is, roughly speaking, g contains only predicates belonging to the common part of the languages L_1 and L_2), we say that $t(g)$ is a *simple translation* of g.

Let us mean by an *inductive L-logic* the Hintikka-type system of inductive logic which is based upon a language L. Suppose that L_1 is translatable into L_2, and that e and e' are complete descriptions of the same sample of individuals within L_1 and L_2, respectively. Then we shall say that the inductive L_1-logic is *equivalent* to the inductive L_2-logic if, for each generalization g in L_1,

$$(10.4) \quad P_1(g/e) = P_2(t(g)/e'),$$

where P_i is the conditional probability measure of the inductive L_i-logic, for $i=1, 2$. (Here we assume, of course, that the values of λ and α are fixed.)

Salmon's interpretation of his criterion CLI_3 as a requirement of ruling out 'interlinguistic contradictions' can now be formulated as follows:

(CLI₇)   If L_1 is translatable into L_2, then the inductive L_1-logic should be equivalent to the inductive L_2-logic.

This condition is not satisfied by Hintikka's systems of inductive logic. Instead, we know that

(CLI₈)   If L_1 and L_2 are intertranslatable, then the inductive L_1-logic is equivalent to the inductive L_2-logic.

A simple consequence of CLI_8 is that

(CLI₉)   If L_1 is translatable into L_2 and L_2 is a sublanguage of L_1, then the inductive L_1-logic is equivalent to the inductive L_2-logic.

That L_2 is a sublanguage of L_1 means here that they have the same individual constants and that the set of primitive predicates of L_2 is a subset of the set primitive predicates of L_1.

Our treatment of the example of Roman bachelors suggests that the violation of the criterion CLI_7 is not as serious matter as Salmon would think. In the first place, Hintikka's system is linguistically invariant in the case of language-shifts which are 'conservative' in the senses of CLI_8 and CLI_9, i.e., in the case of two intertranslatable languages and in the special case where a language is translatable into its sublanguage. In the second place, when a language L_1 is translatable into a language L_2, but not vice versa, the shift from L_1 to L_2, or conversely, is non-trivial or non-conservative. For example, when predicates not belonging to the common part of L_1 and L_2 are exemplified in a sample of individuals, the description of this sample in L_2 is not a translation of the corresponding description in L_1. No equivalence of the corresponding inductive logics should be expected in this case. In the third place, we have seen that an 'interlinguistic contradiction' which is demonstrated in the metalanguage results only from the fact that the metalinguistic meaning postulates are not reflected on the level of the object languages. In inductive logic, similar 'intralinguistic contradictions' arise in the determination of inductive probabilities in the following case: if the primitive predicates of a language L are not logically independent and if, on the level of the object language L, we do not take into account the meaning postulates expressing the definitory connections between these predicates. The difference between these 'intralinguistic' and 'interlinguistic contradictions' lies only in the fact that the metalinguistic meaning postulates connecting two languages L_1 and L_2 may fail to be expressible in either of these languages. However, when we form a new language L_3, as above, which contains all predicates of L_1 and L_2, and compute the inductive probabilities within L_3 conditional on the truth of all meaning postulates MP_0 connecting L_1 and L_2, the 'contradiction' disappears. In Hintikka's system, as it is applied in this work, this procedure of forming the language L_3 has an interesting feature: if L_1 is translatable into L_2, then the inductive L_3-logic is equivalent to the inductive L_2-logic. (This follows from CLI_9 together with the facts that L_3 will be translatable into L_2, and that L_2 is a sublanguage of L_3.) This procedure applies also in the case where only such predicates as those that belong to the common part of the languages L_1 and L_2 are exemplified in a sample. In this case, the description of the sample in L_2 is a simple translation of the corresponding description in L_1. Even here two researchers employing the languages L_1 and L_2 can agree on inductive probabilities if they decide to use the language L_3 which combines their conceptual schemes.

These conclusions illustrate that Salmon's contention that violations of the criterion CLI_7 amount to a *logical* contradiction is not justified. When we take into account, on the level of the object language, the metalinguistic principles on which these 'interlinguistic contradictions' are based, there is no longer a contradiction. But since we cannot express these metalinguistic principles in either of the languages concerned, we are left with the problem of the choice of language. As we have argued above, this choice is not a matter of logic alone, but is dependent upon various pragmatic features. Our results show that the user of the language L_2 into which the language L_1 is translatable (but not vice versa) can make a strong point against the user of L_1, namely, that the inductive logic based on the language which combines L_1 and L_2 is equivalent to *his* inductive logic. Nevertheless, we cannot justify any generally applicable rule for the choice of language on the basis of this result. In certain contexts, the language L_1 may be more appropriate for the purposes of inductive inference than L_2 (or L_3).

This approach to linguistic invariance seems to work well in the much discussed example of colour predicates in different languages. Thus, suppose first that L_1 and L_2 are formalizations of the fragments of English and Finnish, respectively, containing intertranslatable sets of colour predicates. Then, apart from possible notational differences, L_1 would not differ from L_2, and the inductive L_1-logic would be equivalent to the inductive L_2-logic. (Of course, the users of the languages L_1 and L_2 could attribute different values to the parameters λ and α, but this would not amount to a logical contradiction.) Suppose then that L_1 and L_2 are such formalizations of corresponding fragments of two languages having different sets of colour predicates, e.g., of ordinary English and of the language of the Zuni Indians who have a single, primitive name for what we call orange-or-yellow (cf. Hacking, 1969). Then, L_1 permits a finer discrimination of colours than L_2. In other words, L_1 enables its user to make a finer partition of the universe than L_2. In this case, the inductive probabilities within L_1 can differ from the corresponding probabilities within L_2. There is nothing logically contradictory in this, however. The users of L_1 and L_2 apply different conceptual schemes for describing the universe. At least when the evidence contains individuals which are either orange or yellow, the evidence for users of L_1 and L_2 is different. It is then only natural that this difference is also reflected in the corresponding inductive probabilities (see also Hacking, 1969, pp. 27–28, and Salmon, 1963b, p. 53). If only red individuals are observed in the sample, we have a problem of the choice of language. After a sufficient amount of evidence of this kind, the investigators might regard the languages they are using ill-adapted to the problem in question, and delete the non-common (and non-exemplified) predicates from their languages. The Zuni investigator might learn to discriminate between orange and yellow. Or, possibly, the English investigator might agree to use only a part of the conceptual apparatus of his native language, and speak a kind of Subenglish in which the predicates 'orange' and 'yellow' are replaced by a single primitive predicate corresponding to 'orange-or-yellow'. Then his inductive logic would be equivalent to the speaker of the Zuni language (cf. Levi, 1969). Moreover, no logical contradiction can arise in this case.

To summarize our discussion, we have seen that Hintikka's system of inductive logic has reasonable properties of linguistic invariance in that it satisfies the criteria CLI_4, CLI_5, CLI_6, CLI_8, and CLI_9 which are natural interpretations of the general criterion CLI_3. In short, no 'intra-linguistic contradictions' can arise in this system, and for as long as we employ the same conceptual scheme, it does not make any difference to inductive logic how we happen to name the predicates concerned. It seems to us that one cannot expect more linguistic invariance to be built into any system of inductive logic which is sufficiently flexible also to reflect the linguistically variant aspects of scientific inference.

1.3.   In this work, we have been more interested in linguistically variant than linguistically invariant aspects of inductive inference. We have studied situations where a *language-shift* takes place from an observational language $L(\lambda)$ to a richer language $L(\lambda \cup \mu)$, containing theoretical or new predicates. The largely pragmatic question of how the initial language $L(\lambda)$ was chosen has been of no concern to us. Instead, we have focused upon the different problem of accounting for the inductive effects of language-shifts or conceptual change.

It is characteristic of the shift from $L(\lambda)$ to $L(\lambda \cup \mu)$ that $L(\lambda)$ is a sub-language of $L(\lambda \cup \mu)$ and that we can employ the new predicates μ in $L(\lambda \cup \mu)$ to reinterpret the statements originally stated in $L(\lambda)$. The basic step in study of the effects of this shift is that of computing the probabilities of a generalization in $L(\lambda)$ and in $L(\lambda \cup \mu)$. Within Hintikka's system of inductive logic, the latter probability is not directly computable unless certain specific assumptions concerning the nature of the members of μ are made. We have to assume something about their possible use in reporting evidence. The cases we have investigated are those in which they can always be used in reporting evidence (evidential theoretical predicates) and those in which they can never be so used (non-evidential theoretical predicates). For the probability $P(g/e)$ of a generalization g on evidence e in $L(\lambda)$, we have given formulae (3.2) and (4.2). For the probability $P(g/e \& T)$ of g on e and theory T in $L(\lambda \cup \{P\})$, we have given formulae (3.3), (3.17), and (4.3).

The difference

$$P(g/e \& T) - P(g/e)$$

between these two probabilities measures, within our framework, the effect of both the change from $L(\lambda)$ to $L(\lambda \cup \mu)$ *and* the effect of some background information of (either factual or conceptual) assumptions (given in terms of r, b', and partly c'). This difference can be split into a sum of two factors:

(10.5) $P(g/e \ \& \ T_0) - P(g/e)$

and

(10.6) $P(g/e \ \& \ T) - P(g/e \ \& \ T_0),$

where T_0 is either the conjunction of all conceptual assumptions (meaning postulates) connecting the new predicate 'P' with the old predicates λ or, when there are no such assumptions, the null theory obtained by putting $r=0$ and $b'=2b$ in (3.3) and (4.3). The first difference, that is, $P(g/e \ \& \ T_0) - P(g/e)$, measures the pure effect of the change of conceptual framework ('$L(\lambda \cup \mu)$ minus $L(\lambda)$'), while the second difference, $P(g/e \ \& \ T) - P(g/e \ \& \ T_0)$, measures the pure effect of the factual component of T.

On the basis of cases (1)–(6) of Chapter 2, it seems that the difference (10.5) will always be negative for the null theory T_0, so that we have here a case of pure linguistic variance. However, T_0 can here play also the role of meaning postulates to make

$$P(g/e) = P(g/e \ \& \ T_0).$$

Consequently, the present system is quite flexible in the sense that it can take into account both linguistic variance and invariance, depending upon what is desired. Thus, in conservative language-shifts (e.g., when T_0 is an explicit definition of 'P' in terms of λ) our system can be linguistically invariant, while in the cases of more fundamental conceptual changes the relevant probabilities will usually change.

We thus account for total inductive conceptual change jointly by changing the underlying language and by adding new background knowledge. In the next section, this approach is contrasted with some current theories of probability kinematics which, as we shall see, fail to account for the linguistically variant aspects of changes in probabilities.

2. Probability kinematics

2.1.   A topic closely related to linguistic inductive variance as opposed to invariance is *probability kinematics*. Generally speaking, by probability kinematics we mean the study of changes in probabilities.[4] Such changes may be due to several factors. To mention some, we have changes in probabilities due to new empirical or observational information, i.e., *new observational evidence*. This new information may be adopted in the form of accepting as true some statements not previously accepted. For instance, scientific experimentation is supposed to give such new evidence.

Secondly, we may have changes in the probabilities of some statements, changes which are due to changes in the probabilities of some 'basic' statements. The latter changes may be due to *inconclusive observation*, which does not give 'certain' information but which still gives some information encoded only as a change in probabilities (cf. Jeffrey's (1965) example of observation by candlelight).

Next, at least *conceptual revolutions* may result in radical changes in inductive probabilities (cf. Section 1). What we have been studying in this work are in part probability changes due to an enrichment of one's conceptual system (i.e., the shift from $L(\lambda)$ to $L(\lambda \cup \mu)$). However, at the same time, we have studied the effect of new theoretical background assumptions and *theoretical evidence* (stated in $L(\lambda \cup \mu)$) on probabilities of the old statements reinterpreted within $L(\lambda \cup \mu)$.

Let us now briefly review some accounts of probability kinematics to enable us to compare our own approach with them in greater detail. We are familiar with three previous detailed views of probability kinematics. These are the Bayesian conditionalization and its diverse elaborations (see cf. de Finetti, 1937; Carnap, 1962b; and Levi, 1970), Jeffrey's (1965) account, and, thirdly, Lehrer's recent attempt (Lehrer, 1971) which actually applies some previously developed ideas.

In *Bayesian conditionalization* we are dealing with a fixed set of statements (forming a Boolean algebra) for which a fixed probability measure has been defined. Consider now the prior probability $P_0(h)$ of a statement h. The probability of h changes as a function of new evidence e accepted as true, and this change simply occurs according to simple probabilistic conditionalization. The new probability $(P_1(h))$ thus becomes $P_1(h) = = P_0(h/e)$. As our own approach partly incorporates Bayesian conditionalization, we shall defer a more detailed discussion of this view until later.

Jeffrey's model can simply be described as follows. Let $h_1, ..., h_n$ be a set of 'basis' propositions with initial (subjective) probability values $P_0(h_1), ..., P_0(h_n)$ for some agent. Assume now that these probabilities change into $P_1(h_1), ..., P_1(h_n)$. Why such change occurs may be due to the sensory stimulation that our agent receives from his environment. What is essential here in that the agent may not be able (any more than we are) to describe explicitly what caused the change in his probabilities. Therefore, Jeffrey argues, the simple conditionalization model will not work in general as there may be no evidence statement with respect to which to conditionalize. Assume next that the h_i's form a partitioning. Hence, for all $i, j = 1, ..., n$,

$$P_0(h_i) \neq 0$$
$$P_0(h_i \ \& \ h_j) = 0 \quad \text{if} \quad i \neq j$$
$$P_0(h_1 \vee ... \vee h_n) = 1 .$$

Now we make the following *rigidity assumption*: For all $i = 1, ..., n$, and for all k for which $P_0(k)$ exists

$$P_1(k/h_i) = P_0(k/h_i) .$$

That is, even if the probabilities of the h_i's change, and even if the probabilities of such k's will change as a function of the change in the probabilities of the h_i's (see below (10.7)), the k's still retain their conditional probabilities relative to the truth of the h_i's.

Now, if the above assumptions are satisfied, then, on the basis of the elementary probability calculus, we derive

(10.7) $P_1(k) = P_1(h_1) P_0(k/h_1) + \cdots + P_1(h_n) P_0(k/h_n).$

Formula (10.7) thus describes how the probabilities of statements other than the h_i's change from their P_0-values to P_1-values, given the changes in the probabilities of the h_i's (and the partitioning and rigidity assumptions).

Let us leave later our comments on Jeffrey's model and go on to Lehrer's (1971) view. What Lehrer has to say on inductive change is not much more than an application of Jeffrey's model on the one hand and of Kemeny's and Carnap's views on the other hand. Furthermore, the details of Lehrer's account are bound to his probabilistic theory of evidence, which we find unacceptable.[5] We can accordingly be very brief here, and take up only a couple of generally interesting issues.

Lehrer distinguishes between two types of conceptual change and tries to account for them in terms of inductive change (probability kinematics) (cf. Lehrer, 1971, pp. 215–220). The first type of conceptual change is the result of considering a problem in an altogether new way. (To conceive of constant motion as natural without any first motion might serve as an example here.) In terms of inductive change, this means a radical shift in the probabilities of some relevant hypothesis (corresponding to Jeffrey's basic statements). The changes in the probabilities of other statements may then be obtained by means of Jeffrey's formula (10.7). Lehrer accepts this solution. But as a criticism of this, one may of course say that (1) this account provides no philosophical explanation of the radical shift in the probabilities of the basic statements and that (2) there is no guarantee that the rigidity assumption is true in radical conceptual shifts.

The second kind of conceptual shift is a profound one. It involves a change in the semantical status of some statement, i.e., a change from contingent to noncontingent, or conversely. Suppose, for instance, a statement h changes from being contingent to being logically true. We may then simply use conditionalization to obtain the new probability for any k:

$$P_1(k) = P_0(k/h),$$

(It is thus assumed that only h is directly effected by conceptual change.) More generally, we may here use the meaning postulate technique of Carnap and Kemeny: assume that conceptual changes in semantical status can be encoded by means of some meaning postulated stated within the language in question, and conditionalize the probabilities to these meaning postulates. But this kind of conditionalization may prove to be too simple, Lehrer argues. Lehrer sketches another possible solution. He operates within a framework within which the basic statements form a partition. In such a situation, we may be able to say, for instance, when one member of the partition becomes logically or semantically contradictory and receives a zero new probability. We may then discuss and develop various methods for redistributing the probability mass of that member to other members (cf. Lehrer, 1971, p. 218).

2.2. One important criticism applies to all of the earlier accounts of probability kinematics: they do not treat conceptual change in which totally *new* concepts are introduced to the framework (cf. Suppes' (1966) criticism). However, our approach in this work does in fact attempt to do this (though admittedly within a simplified framework). A conceptual change is reflected in the shift of a language $L(\lambda)$ into a new language $L(\eta)$ in which the members of η include novel concepts. The case which has been the main subject of our interest in this work is that with $\eta = \lambda \cup \mu$, λ being a set of observational concepts and μ a set of theoretical concepts. The essential point within our approach is that we have used the new and richer framework $L(\lambda \cup \mu)$ to *reinterpret* or redescribe the generalizations g and the evidence e originally formulated within $L(\lambda)$.

We have assumed that conceptual shifts, i.e., new ways of conceiving of problems, normally involve both the addition of new concepts and of new background assumptions and information T (statable only by means of the novel concepts). However, these two factors can be kept conceptually and, more specifically, inductively distinct.

Conceptual changes have inductive consequences – they change the inductive probabilities P_0 defined for $L(\lambda)$ in to new probabilities P_1 defined for $L(\lambda \cup \mu)$. Thus we have two (and not only one) basic problems for probability kinematics:

(1) How do the probabilities P_0 and P_1 relate? (2) What is the effect of new information on inductive probabilities in general (be these probabilities defined for $L(\lambda)$ or for $L(\lambda \cup \mu)$)? In answering the first question, we have mainly been discussing the difference

$$(10.8) \quad P_1(g/e \ \& \ T) - P_0(g/e),$$

by virtue of its great methodological interest. Now, this difference (10.8) measures the inductive effect of both the change from $L(\lambda)$ to $L(\lambda \cup \mu)$ *and* the effect of the background assumption T (given in terms of r, b', and partly c'). To distinguish these factors, we consider the following two differences:

$$(10.9) \quad P_1(g/e \ \& \ T) - P_1(g/e \ \& \ T_0)$$

$$(10.10) \quad P_1(g/e \ \& \ T_0) - P_0(g/e).$$

Difference (10.9) measures the effect of the factual information contained in T. T_0 is as in Section 10.1. Then (10.10) measures the pure inductive effect resulting from the change of conceptual framework ('$L(\lambda \cup \mu)$ minus $L(\lambda)$'). In Section 2, and earlier, we discussed the relations between $P_1(g/e \ \& \ T)$, $P_1(g/e \ \& \ T_0)$ and $P_0(g/e)$ in detail. We noted that our approach is rather flexible in questions of linguistic invariance as opposed to variance, as T_0 may be taken to contain meaning postulates to adjust the difference (10.10).

Inductive probabilities within Hintikka's system depend upon two extra-logical parameters, α and λ, which can be interpreted as reflecting the regularity or orderliness of the world (cf. Chapter 2). The parameters α and λ thus represent presuppositions of induction. These parameters are interpretable objectivistically in the sense that they are conceived to indicate the true degree of regularity of the world. This is, of course, an idealized notion. When applying inductive logic, a scientist has to estimate the values of these parameters. In doing this, he relies both on his empirical evidence and on his theoretical background information and assumptions. Among the latter are included the research canons of the community of scientists to which he belongs.

In principle, the values of α and λ may depend on various factors. Two interesting cases in this respect are the following. Conceptual change from $L(\lambda)$ to $L(\lambda \cup \mu)$ may sometimes go together with a change in the values of α and λ. Such a change may also occur, within one and the same conceptual framework, in going from one evidential situation to another. Both of the cases are of course relevant to probability kinematics, as they entail changes in probability measures.

Situations in which a change in probabilities results from a change in the values of the parameters α and λ do not fit together with the simple Bayesian conditionalization model. In this book, we have restricted our treatment to inductive situations with fixed parameter values. For this reason, we have been able partly to incorporate the conditionalization model within our framework. In fact, we have operated as if we had access to the true parameter values, which are objective characteristics of the world and not dependent upon any scientist's estimates of these characteristics. Another way of interpreting the situation is that of treating the values of α and λ as estimates of the true values, and adding that these estimates remain fixed.

While we do not know of any previous attempts seriously concerned with treatment of problem (1), all of the accounts reviewed have related to some aspect of the kinematic problem (2). Our own view of problem (1) is logically compatible with any of the previous partial solutions to (2). It may consequently be taken to complement rather than to replace these accounts. Nevertheless, we wish to indicate our philosophical standpoint concerning the best acceptable approach to (2). Consider first Jeffrey-type solutions. Within our framework, for instance, the constituents of both $L(\lambda)$ and, alternatively, of $L(\lambda \cup \mu)$ qualify as the statements h_i forming an inductive partitioning. The question is whether (a) there are cases in which the evidence cannot be explicitly formulated and justified and in which (b) the rigidity assumption holds true. If (a) is false, Jeffrey's prime reason for rejecting the simple conditionalization model is not acceptable. Even if (a) were true, Jeffrey's model still presupposes (b).

To begin with (b), it seems justifiable for many practical decision situations, with which Jeffrey is mainly concerned. However, we do not see why the rigidity assumptions should hold good in those *scientific* cases in which (a) would be true. However, maybe this is just attributable to our belief in the falsity of (a). Why then do we take (a) to be false, for interesting scientific cases? This is because we think of scientific theorizing and experimentation as prime examples of reasoned deliberate action. We wish to claim that these reasons can be given in explicit terms as required, for instance, by the conditionalization model. (The reasons may consist, say, in citing evidence, observational or theoretical.) In all, our rejection of the above assumption (a) for normal scientific cases, together with our previous criticism that Jeffrey's model does not give any philosophical explanation of probability shifts, provides us with good reasons not to accept Jeffrey's model as a *general* answer to the kinematic problem (2). (This is not to say that Jeffrey's model might not adequately account for changes in degrees of belief in many situations of ordinary life.)

Our approach to probability kinematics, then, is briefly this. We accept a *qualified* version of Bayesian conditionalization as an answer to problem (2), and the above treatment of conceptual change (with fixed parameter values) as an answer to (1).

In regard to our version of the conditionalization method, the following additional remarks may be made. In this book, we have been advocating a hypothetico-inductive method of theorizing. What is hypothetical about it is at least the (truth of the) background theory (T). Nevertheless, we are methodological empiricists accepting fallibilism: T should be empirically criticizable, indirectly, if not directly, and in a piecemeal fashion, if not *in toto*. Consequently, T must in principle be explicitly stated (as of course all the other assumptions and statements as well).[6] Hence Jeffrey's assumption (a) is normally false for scientific cases.

Another relevant problem which may have led Jeffrey (and certainly Lehrer, 1971) to reject the conditionalization model is this. Within the conditionalization model, evidence will have the probability one, as $P_1(e) = P_0(e/e) = 1$. Jeffrey and Lehrer seem to think that this means certainty and that certainty means incorrigibility. We accept that $P_1(e) = 1$ explicates certainty, but we do not accept that it explicates incorrigibility (cf. Levi, 1970). This can be argued as follows. We are inclined to interpret our inductive probabilities objectivistically. Thus inductive probability statements are for us *factual* (even if not always strictly empirical) statements. (Recall what was earlier said about the extralogical nature of the inductive parameters α and λ.) On the other hand, every application of inductive logic to real cases of inductive inference can take place only by means of the scientist's estimates of these objective probabilities. Hence these estimates can be considered as credence-probabilities. In terms of these credencies, evidence statements certainly can have the probability one. As evidence statements e are corrigible, also the credence probabilities P_1 (such that $P_1(h) = P_0(h/e)$) are corrigible. If e is now retracted, then also the measure P_1 (for which $P_1(e) = 1$) is retracted. Thus they are retractable and corrigible. (Cf. Levi's (1970) account, which we basically accept.) In all, applied to nonretracted evidence, simple conditionalization seems to work as a solution to problem (2). (In cases where previously accepted evidence is later retracted, conditionalization has to be reapplied in the obvious way to nonretracted evidence.)

As a final comment on probability kinematics we wish to emphasize once again that no solution to problem (2) alone suffices, but that problem (1) also has to be accounted for. Our approach is a small step in this direction.

3. GOODMAN'S NEW RIDDLE OF INDUCTION

Even if inductive logic alone may not solve the philosophical problems of induction it may at least help in clarifying the various aspects of such problems, and thus to sharpen various philosophical solutions. This has been our viewpoint in this book when attacking difficult philosophical problems. When discussing such deep problems as Goodman's paradox we emphasize more than elsewhere that inductive logic and technical logical work is no substitute for philosophical thinking but at best an auxiliary tool.

Let us now go to discuss Goodman's famous 'new riddle of induction'. It can be described by means of an example – slightly different from Goodman's (1955) – as follows. Assume that we are investigating the color of emeralds. We have observed n emeralds by now and found all of them to be green. Now define a new predicate 'grue' as follows. Let '$R(x)$' stand for 'x is grue', '$G(x)$' for 'x is green', '$E(x)$' for 'x is an emerald' and '$P(x)$' for 'x is examined before 2000 A.D.'. Then we define

$$(10.11) \quad (x)\,(R(x) \equiv ((P(x) \supset G(x)) \ \& \ (\sim P(x) \supset \ \sim G(x)))).$$

In other words, an object is defined to be grue just in case it is and found to be green if examined before 2000 A.D. and found to be non-green if examined after 2000 A.D..

Now consider the following two hypotheses:

$$(10.12) \quad (x)\,(E(x) \supset G(x))$$

$$(10.13) \quad (x)\,(E(x) \supset R(x)).$$

Hypothesis (10.12) states that all emeralds are green whereas (10.13) tells us that they are grue. Both hypotheses have the same syntactical form and they also have the same evidence (recall that green objects examined before 2000 A.D. are also grue objects). Still, intuitively (10.12) is much better supported in terms of this evidence than (10.13), even if (10.12) and (10.13) receive the same inductive probability in all purely logical (syntactical) inductive logics. How do we explain this discrepancy?

As is well known, Goodman himself has tried to account for the better projectibility of (10.12) *pragmatically* in terms of the previous inductive *entrenchment* of the individual predicates occurring in these hypotheses. We shall not here discuss this account, nor shall we discuss most of the other attempts to solve Goodman's paradox. Instead, we formulate some strong criteria of adequacy for a solution, and then go on to discuss the paradox under these requirements and in the light of our earlier results in this work. Our starting point is very close to that of Hesse's (1969) and the reader is referred to her paper for a justification of the criteria of adequacy for a solution.

Predicate 'P' in our above example refers to *time*. That this is inessential to the paradox has been convincingly argued by various authors (cf. Scheffler, 1963). Furthermore, it is inessential that 'P' is given in pragmatic terms rather than objectivistically. Still another inessential assumption (sometimes made) is that emeralds change colour if (10.13) is true (cf. Hesse, 1969). In fact, we shall assume that both Mr Green – the advocate of (10.12) – and Mr Grue – the advocate of (10.13) – will accept that emeralds retain their colour after 2000 A.D.

What is philosophically essential is that a solution of Goodman's paradox should satisfy the following criteria of adequacy (cf. Hesse, 1969a):

(C1) The predictions concerning the color of emeralds after 2000 A.D. by Mr Green and Mr Grue should be genuinely (factually, and not merely verbally) different. Furthermore, the languages of Mr Green and Mr Grue should be intertranslatable in some strong sense so that Mr Green and Mr Grue can agree that their predictions are factually different and incompatible.

(C2) Even if the hypotheses (10.12) and (10.13) (and the predictions concerning the colour of emeralds examined after 2000 A.D.) should be inductively asymmetric, the paradox should be stated and solved without introducing needless linguistic asymmetries between the problematic predicates.

Criterion (C2) is somewhat vaguely stated. Despite that, it clearly excludes a number of attempted solutions. For instance, it excludes all solutions which claim that 'grue' is somehow intrinsically more complex than 'green' merely because it can be defined in terms of 'green' and 'P'. Let us rewrite (10.11) as

(10.14) $(x) (R(x) \equiv (G(x) \equiv P(x)))$.

Now we see immediately that given 'R' and 'P' we can define 'green' symmetrically by

$$(10.15) \quad (x)\,(G(x) \equiv (R(x) \equiv P(x))).$$

In fact, if we can give a satisfactory explanation of why to begin with 'green' rather than with 'grue', we have philosophically solved Goodman's paradox.

The situation can now be viewed in general terms as follows. We are dealing with two theoreticians, Mr Green and Mr Grue, who employ somewhat different conceptual frameworks, and have adopted different theories about the world. Their frameworks are (or should be) inductively different. However, they are able to communicate with each other, so that their frameworks are semantically commensurable and translatable. Let us now go into some more detail of the situation by discussing the emerald-example (which may, however, be somewhat misleading as no scientifically-interesting theoretical predicates are involved).

Mr Green's vocabulary initially consists of $\lambda_G = \{E, G\}$. He adopts (10.12), call it g_G, as his general assumption concerning the world. Let his estimate of the degree of regularity of the world as required by Hintikka's inductive logic be α_G. We may assume that his evidence e_G consists of n observed individuals, which exemplify the cells EG, $\bar{E}G$, and $\bar{E}\bar{G}$ (cf. Figure 10.1 below). We can then compute the probability $P(g_G/e_G) = P_G$. Analogously, Mr Grue's initial vocabulary is $\lambda_R = \{E, R\}$. His evidence e_R is assumed to consist of the same individuals as Mr Green's but described by means of 'grue'. Given Mr Grue's estimate α_R of the degree of regularity of the world (with respect to λ_R) we have fixed the probability $P(g_R/e_R) = P_R$.

It is now easy to see that, given our assumptions,

$$(10.16) \quad P_G > P_R \text{ iff } \alpha_G < \alpha_R.$$

It seems reasonable to claim that $\alpha_G < \alpha_R$ for the following reason. We have made the strong assumption that full communication is possible between Mr Green and Mr Grue. This may be explicated by saying that their languages L_G and L_R are intertranslatable in the sense of Section 1.

Upon what is such full communicability and intertranslatability based? It seems that it is partly based on Mr Green's and Mr Grue's sharing of certain general linguistic principles (principles of 'understanding') and partly on their correlating language with the external world by similar means. It suffices here to consider this latter ostensive factor, that is, the objective criteria for the application of primitive descriptive predicates. If there is any economy at all in man's conceptual system designed for gathering information from the external world, we can say this: our conceptual system should reflect the general regularity and orderliness of the world. Thus, for instance, if jewels tend to have *simple* permanent properties, our conceptual system should reflect this. In the case of colour-predicates such as 'green' as against 'grue', we may employ objective meaning criteria, such as wavelengths, for their application. Now, of any objective criteria we (or any normal human being) can conceive of, 'green' will need only one criterion (or one set of criteria) whereas 'grue' requires two (first that used for 'green' and then another one after 2000 A.D.).[8] This we take to reflect the underlying estimates of the regularity of the world. Hence we always have $\alpha_G < \alpha_R$, which should then be granted by both Mr Green and Mr Grue. It follows that $P_G > P_R$. Therefore we may say that the *degree of lawlikeness* of the conceptual system L_G (and hence of g_G) is greater than that of L_R (and hence g_R). (Hintikka (1969b) has suggested that the parameter α can be taken as an indicator of the degree of lawlikeness of a conceptual system.)

This may be regarded as one acceptable way of solving Goodman's paradox as our criteria (C1) and (C2) will now become satisfied. To see this, the requirement of intertranslatability still needs some clarification.

It seems appropriate to understand the communicability condition (C1) to mean intertranslatability in the strong sense of Section 1. At least this assumption seems to have been implicitly or explicitly adopted by philosophers discussing Goodman's paradox (cf. Hesse, 1969a). However, one qualification is needed. It is allowed that to establish intertranslatability the vocabularies of both Mr Green and Mr Grue may have to be enriched by some new common 'auxiliary' predicates. (This may be motivated by claiming that communication cannot always, for practical or for philosophical reasons, be achieved by means of ostension only.) In our example we add $\mu = \{P\}$ both to λ_G and λ_R. Thus we create two new languages L_{GP} and L_{RP}.

The present situation viewed from the viewpoint L_{GP} can be illustrated by the following figure:

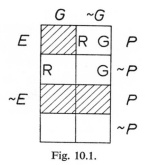

Fig. 10.1.

In this figure the shaded cells represent our evidence. Of the blank cells those marked G are empty due to g_G, while those marked R are empty by g_R.

The resulting languages L_{GP} and L_{RP} can be shown to be intertranslatable in the metalanguage if metapostulates corresponding to (10.14) and (10.15) are adopted as meaning postulates. Their intertranslatability is made more explicit by constructing a language L_{GRP} with 'E', 'G', 'R', and 'P' as its primitive predicates, satisfying (10.14) and (10.15).

We know from Section 1 that the languages L_{GP}, L_{RP}, and L_{GRP} are all *inductively equivalent*, provided α_{GP}, α_{RP}, and α_{GRP} have identical values. Hence g_G and g_R obtain the same relative probabilities in L_{GP}, L_{RP}, and L_{GRP}. In this case also the translations $t(g_G) = (x) (E(x) \supset (R(x) \equiv P(x)))$ and $t(g_R) = (x) (E(x) \supset (G(x) \equiv P(x)))$ into L_{RP} and L_{GP}, respectively, get both this very same probability value, which is, by the way, smaller than the value in either L_G or L_R (for the same α-values).

Is there any good reason to suppose that $\alpha_{GP} = \alpha_{RP} = \alpha_{GRP}$, given that $\alpha_G < \alpha_R$? It seems that this problem is strongly context-dependent. In some fields of science, the addition of new predicates and thus conceptual change might conceivably effect the scientist's estimates of α. For instance, we might think that we have $\alpha_{GP} < \alpha_{RP}$ but that both Mr Green and Mr Grue agree on their value of α_{GRP} (perhaps they make it equal to α_{RP}) after realizing the symmetry expressed by (10.14) and (10.15). If they could always find a common extended language and agree on its α-value, the paradox would vanish and we should accept the same degrees of confirmation for g_G and g_R. According to our earlier arguments, however, this need not be the case if Mr Grue employs very 'perverse' predicates. It may be that Mr Grue can go far in preserving the symmetry required by the strong intertranslatability, even if he may finally have to give up (this need not contradict our criterion (C1) as it does not imply *strong* inter-translatability by means of explicit definitions). But even if Mr Grue could thus preserve the purely linguistic symmetry the objective criteria of application for his conceptual system would get too complicated in our actual world, provided that he and Mr Green share some basic research interests which direct the choice of objective application criteria for predicates.

Our main argument has so far rested on the assumption that Mr Green and Mr Grue are likely to differ, and agree to differ, in their estimates of α (at least for L_G and L_R and probably for the richer languages as well). The second major factor in solving the paradox has been the problem of whether Mr Green and Mr Grue are able to communicate in the strong sense of intertranslatability. If this assumption cannot be satisfied, Mr Green's system is inductively better without further consideration.

There is still another relevant factor affecting the paradox. Even if Mr Green and Mr Grue would always agree in their α-estimates and even if their languages were intertranslatable, their differing background assumptions might make their probabilities for g_G and g_R different. That is, as we have seen earlier, often $P(g_G/e_G \,\&\, T_G) > P(g_R/e_R \,\&\, T_R)$ even if $P(g_G/e_G) = P(g_R/e_R)$ and even if $T_G = T_R$. What could such background knowledge be like? First of all, we have to note that T_G and T_R may necessitate the introduction of new theoretical concepts and thus conceptual enrichment. However, as we have assumed that Mr Green and Mr Grue agree on the proposition that emeralds do not change colour (in reality), T_G and T_R must agree in their purely factual content. For instance, T_G and T_R may contain as their common factual part a physical theory of wavelengths. What is essential to the controversy between Mr Green and Mr Grue is that T_G and T_R also express purely conceptual or linguistic assumptions, that is, assumptions about the conceptual systems of Mr Green and Mr Grue. To avoid inessential asymmetries in the sense of our condition (C2), we below assume that Mr Green and Mr Grue can agree on such assumptions, so that $T_G = T_R = T$. Before we proceed to some specifics, it needs to be emphasized that in general such a theory T will contain a second-order theory which gives properties to concepts (predicates). Consequently, our framework is too poor fully to handle such a theory. However, theory T

may have first-order consequences which suffice for our purposes. We proceed to give an example of this.

Let us introduce a new predicate 'Q' as follows:

$Q(x)$ = the objective criterion for the colour of x is not the same as before 2000 A.D.

Our background theory then amounts to saying that for 'green' we always use the same objective semantic criterion, whereas the criterion for 'grue' changes at 2000 A.D. As before, we assume this is accepted both by Mr Green and Mr Grue. Our background theory can now be explicated to have the following conjunction, called T, as its first-order consequence:

(10.17) $(x)\,((E(x)\ \&\ G(x)\ \&\ \sim P(x)) \supset\ \sim Q(x))$ &

$(x)\,((E(x)\ \&\ \sim G(x)\ \&\ \sim P(x)) \supset\ \sim Q(x))$ &

$(x)\,((E(x)\ \&\ R(x)\ \&\ \sim P(x)) \supset Q(x))$.

This situation can be illustrated by means of the following figure, to be compared with Figure 10.1.

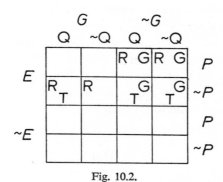

Fig. 10.2.

In this figure, the cells marked T are those specified to be empty by T. We assume that Mr Green and Mr Grue have accepted the same evidence (with respect to Q) and that otherwise the situation is the same as in Figure 10.1.

Within the language L_{GPQ} (and L_{RPQ}), we then have (according to our earlier results)

$$P(g_G/T \ \& \ e) > P(g_R/T \ \& \ e)$$

as T makes g_G and g_R asymmetrical in Figure 10.2. That is, in the light of T g_R makes empty one more cell than g_G. This shows that at least within some cases new (theoretical) concepts may constitute the solution of Goodman's paradox, even within such simple frameworks as ours.

Our discussion of Goodman's paradox does not account for all the relevant factors, even within our interpretation of the constraints (C1) and (C2), still less for other problems that arise from complete rejection of one or both of them.

Nevertheless, within these constraints we have presented two possibilities of solving Goodman's paradox, the first by means of changing the α-values, the second by introducing a background theory T employing new concepts. Intuitively, both solutions rely on the differences in the objective application criteria of the disputed predicates. In the first solution, the inductive differences between 'green' and 'grue' are based upon a *presupposition of induction* expressed by α. In the second solution, we tried to account for these differences in terms of *background theory* expressed in a language which is a common extension of both Mr Green's and Mr Grue's original language.

Within Hintikka-type inductive logics, Goodman's paradox has been discussed by Hintikka (1969a) and by Pietarinen (1972); they have devised technical methods for creating asymmetries between primitive predicates. However, they do not give such a philosophical justification for these asymmetries that our criterion (C2) is satisfied.

NOTES

[1] Lakatos suggests that the essential part of 'language-planning' is a 'mere by-product of scientific theorizing': one should choose the language of the 'most advanced, best corroborated theory'. In this sense, Lakatos contends, the problem of the choice of language is reducible to the Popperian problem of corroboration of theories (cf. Lakatos, 1968, p. 364). We agree with Lakatos in admitting that the stage of scientific theorizing at a given time may be highly relevant to the choice of language. However, theoretical knowledge expressed in scientific theories seems to be only *one* relevant factor in the evaluation of conceptual systems, and in the choice of certain particular system in a given context. Consequently, a theory of corroboration alone does not give a satisfactory general solution to the problem of the choice of language.

[2] One way of approaching this question is that of drawing a sharp distinction between a *conceptual system* and its diverse expressions in a *linguistic* form. Using the notion of intertranslatability between languages (cf. below), conceptual systems might be taken to correspond to equivalence classes of intertranslatable languages. Differences between intertranslatable languages can be regarded as arbitrary, while differences between languages which are expressions of different conceptual systems are essential or non-arbitrary. In this sense, it is plausible to require that inductive logic is variant with respect to conceptual systems, and invariant with respect to intertranslatable languages (cf. criterion CLI[8] below).

[3] This definition of translatability is sufficient for our purposes here. It can be easily generalized to more complex situations. For example, if the primitive predicates in λ_1 and λ_2, respectively, are not logically independent, the corresponding meaning postulates should be added into the set MP. Note that if the set MP_0 is also allowed to include non-analytical or factual statements accepted as true, then a notion of *factual translatability* is obtained. The growth of scientific knowledge often has the consequence of showing the factual translatability of one linguistic system into another.

[4] These probabilities are to be understood as objective inductive probabilities or subjective probabilities ('credence functions' and the like). One's probability kinematics obviously depends quite strongly upon one's probability interpretation.

[5] Lehrer tries to define evidence in terms of the subjective probabilities of competing statements (see Lehrer, 1971, pp. 203 and 213). Furthermore, he requires that no contingent statement will have a P_0- or P_1-probability of one or zero (p. 208). However, we think that evidence cannot be characterized probabilistically in the manner of Lehrer. (Cf. also Niiniluoto's, 1971, criticism of Lehrer's rule of acceptance RIE (p. 213).) To be sure, within our account a few conclusions follow concerning the probabilities of evidence statements. One such conclusion is that evidence statements have the P_1-probability one.

[6] We are willing to accept Levi's (1970) approach to the problem of total evidence in our characterization of T. (In fact, T & e will normally represent our total evidence.) In some cases, T may characterizable only indirectly in terms of its inductive consequences.

[7] These objective probabilities still are not measures of degrees of (informative) truth. Rather our measure $corr_4$ (see Chapter 8) can be considered to measure verisimilitude or closeness to truth.

[8] This point is better appreciated if we define, as Goodman did, that emeralds are grue if and only if they are green before a certain point of time and blue after that. It should be emphasized that we speak here of *objective* meaning criteria as against such non-objective criteria as 'pictorial representation' employed in Barker and Achinstein (1960). Note that it is logically possible, though not easily conceivable, that Mr Green and Mr Grue could agree on the objective criteria for 'green' and 'grue' such that 'grue' has only one but 'green' two or more criteria. The extent to which we have philosophically solved Goodman's paradox ultimately reduces to this problem.

CHAPTER 11

TOWARDS A NON-INDUCTIVIST LOGIC
OF INDUCTION

According to the old Aristotelian dictum, one can have knowledge only of the general. In the spirit of this epistemological tradition, philosophers have often emphasized that the proper subject-matter of science includes only those aspects of the world which are, in some sense, invariant, uniform, or regular.[1] While the Ancient thinkers were apt to find these uniformities in the unchanging 'forms' or 'essences' of particular things, the founders of modern natural science sought them in the invariant relations or functional dependences between things and events.[2] A common tenet of these views is that science proper studies invariable uniformities which can be expressed by *universal laws*. It is also generally thought that these laws are explained by more comprehensive and general *theories* employing characteristic theoretical concepts and postulates. Thus, all genuine scientific knowledge, as it is expressed in theories and laws, should be stated in universal form.

On the other hand, science – even if it seeks knowledge of the general – is connected with particular things and events. In the first place, since we always perceive only particular things, observational evidence for theories and laws is always stated in singular or existential form only. However, because of the theoretical and general character of scientific theories and laws, scientific knowledge 'transcends' or goes beyond the observational evidence in a double sense. In the second place, in the general nature of scientific theories and laws lies the important feature that they allow us to make predictions about future or yet unobserved events. This power of general theories to yield reliable singular predictions is also the basis of applied science or technology.

The traditional problem of induction emerges from this situation. Scientific knowledge consists of general theories and laws. But how do we arrive at them? How do we justify the claim that we have found theories which are close to truth? How do we account for the rationality of the trust in theories in their present and future applications? Or how do we account for the success and reliability of the singular predictions derivable from theories?

The provision of an answer to these problems has been a serious challenge to the theorists of scientific inference. In this chapter, we discuss some answers which have been put forward as (partial) solutions to these problems, and thereby try to indicate the light that may be shed by our approach and results in this book on such questions as the nature of inductive inference, and its role in science, and how the role of theories can be accounted for in a logic of induction. The frequent failure of attempts by the theorists of induction is not a result of the impossibility of developing a satisfactory system of inductive logic, even though some critics have hastily drawn this conclusion. It seems to us that it rather arises from the narrow empiricist and 'inductivist' philosophical outlook of many theorist of induction. Consequently, one can agree with Mario Bunge's wish that "a noninductivist logic of induction should be welcome" (cf. Bunge, 1963, p. 152). Our development of Hintikka's system of inductive logic is a small but resolute step towards this goal.

1. DEDUCTIVISM AND INDUCTIVISM

1.1. The nature of induction and its role in scientific inquiry have been the subjects of lively philosophical debate ever since the days of Aristotle. 'Deductivists' have maintained that scientific inference is always deductive, while 'inductivists' have emphasized its predominantly inductive nature. However, since the term 'induction' has been used in various different senses during the course of this debate, there are also many different views which have been (or can be) called 'inductivist'.

To separate some important senses of 'inductivism', let us first quote Bunge's definition (cf. Bunge, 1963, p. 148):

According to inductivism, empirical knowledge (a) is obtained by inductive inference alone, (b) is tested only by enumerative induction, (c) is more reliable as it is closer to experience (epistemologically simpler), (d) is more acceptable as it is more probable, and consequently (e) its logic – inductive logic – is an application of the calculus of probability.

As one cannot expect to find many theorists who would accept *all* the conditions (a)–(e), below we define five different senses of inductivism on the basis of somewhat sharpened forms of these conditions.

According to *inductivism*$_1$, all factual (non-analytic) knowledge is obtained either through observation or through inductive inference from statements reporting observations. As an example of inductivism$_1$ doctrine, one can mention Francis Bacon's theory of scientific inference, in which one rises by successive steps from particulars of simple sensuous perception to the most general and abstract axioms. This gradual process is supposed to take place according to the canons of eliminative induction without recourse to deduction.

According to *inductivism$_2$*, scientific inquiry proceeds through the successive steps of observation and classification of facts, inductive generalization from them, and further testing of the generalizations. Here deduction may have an important function in the development of consequences of generalizations obtained through induction. In inductivism$_2$, induction is thought of in an enumerative sense as an inference from particular cases to a general conclusion, which takes place without any recourse to preliminary hypotheses and to background knowledge or presuppositions. In other words, scientific laws and theories are supposed to be derived inductively from observed facts, so that induction is a *method of discovery* of laws and theories. Thus, inductivism$_2$ corresponds to what Hempel (1966) calls "the narrow inductivist conception of scientific inquiry."

It is now commonplace to claim that inductivism$_2$, or the doctrine of inductive generalization (enumerative induction) as a method for the discovery of theories is untenable. Thus, it has been claimed that scientific theories are "happy guesses" (cf. Whewell, 1847) or "conjectures" (cf. Popper, 1963) which are "not *derived* from observed facts, but *invented* in order to account for them" (cf. Hempel, 1966, p. 15). The invention of theories is usually preceded by an *analysis* of the relevant factors of the experimental situation[3] (cf. Hintikka, 1969c) and the introduction of *new concepts* to account for observed facts. Theories employing new concepts are then used to account for, to explain, to reinterpret, and to correct the observations (cf., *inter alia*, Feyerabend, 1965). Thus, observation in science takes place relative to hypotheses, theoretical background knowledge and presuppositions. All this implies that inductivism$_2$ is a seriously mistaken doctrine.

It should be noted that all philosophers who have used the name 'induction' for the inference *to* laws and theories are not inductivists$_2$. For example, Aristotle defines *epagoge* (which is usually translated as 'induction') as the way of coming to know the first principles of a science. However, *epagoge* is not the same as inductive generalization from observations in an enumerative sense, but is a process inseparable from concept formation or the establishing of definitions in science (cf. Hintikka, 1972). Francis Bacon says that induction by simple enumeration is "childish" (cf. Bacon, 1620, I, 105), and William Whewell (1847) emphasizes that induction is a process where facts are 'colligated' by a new concept.[4] Therefore, when inductivism$_2$ is rejected, it seems advisable to use some name other than 'induction' for the process of inference to laws and theories. For example, Peirce's term *'abduction'* might be used for this purpose.

It can be debated whether abduction is only a process of 'free invention' or whether there are rules of abduction, that is, a 'logic of discovery' (cf. for example, Hanson, 1961). In any case, abduction is not simply the same as enumerative or eliminative induction. Therefore, the impossibility of a logic of abduction would not imply the impossibility of a logic of induction (cf. von Wright, 1951).

There are various different anti-inductivist$_2$ doctrines which claim that scientific laws and theories are not obtained by inductive inference from antecedently observed data, but rather by inventing hypotheses or conjectures, and by subjecting these to empirical tests. These hypotheses are then evaluated and accepted on the basis of observed test data which do not deductively entail them, but only give them inconclusive 'inductive support'. According to these views, induction is not a method of discovery but a *method of validation* or justification of hypotheses, and scientific inquiry is "inductive in a wider sense" (Hempel, 1966, p. 18).

According to *inductivism$_3$*, the validation of hypotheses is based on enumerative induction. This doctrine corresponds to some simple formulations of the *method of hypothesis* or the *hypothetico-deductive method* according to which hypothesis are confirmed or supported by their positive instances or by verifying their deductive consequences. Then those non-falsified hypotheses that are sufficiently confirmed are tentatively accepted (cf. e.g. the discussions in Barker, 1957; Salmon, 1966, and in Section 2 below).

Another view closely connected with the method of hypothesis is *inductivism$_4$*. According to this, one should choose the 'simplest' hypothesis compatible with observed data, because it is known by induction that this simplest hypothesis is the one which is most probably true (cf. the discussions in Kemeny, 1953; Goodman, 1959; Ackermann, 1961; Barker, 1961b; Bunge, 1961; and Quine, 1963).

The doctrine of *inductivism₅* takes as its starting point the idea of applying 'inverse probability' to the inductive problem of justifying hypotheses (cf. Jevons, 1877). This idea was developed into a probabilistic theory of induction by the Cambridge School (Johnson, Broad, Keynes, Nicod, Ramsey, Jeffreys), Carnap and Reichenbach. According to inductivism₅, the posterior inductive probability of a hypothesis on a given evidence can be interpreted as the degree to which this evidence supports or confirms this hypothesis. The inductive probability of a hypothesis may depend on the number and variety of its instances exemplified in the evidence. (Here inductivism₅ differs from inductivism₃.) Usually, it is also thought that a high degree of confirmation is necessary and sufficient for the acceptance of a hypothesis. Thus, inductivism₅ corresponds to what in Chapter 8 we called theories of support which are probabilistic in the narrow sense. (See also Section 3 below.)

The inductivist doctrines defined above are intimately connected with a narrow form of the empiricist tradition in the philosophy of science. In them, it is thought that scientific knowledge is either obtained by inductive inference from observations, or supported inductively by observational test data. Observation itself is unproblematically seen as neutral *vis-à-vis* theories or theoretical presuppositions. The greater the epistemological simplicity of scientific knowledge is (in the sense of Bunge, 1963), that is, the closer it remains to one's sense experience, and the more it avoids the employment of theoretical terms referring to unobservables, the more reliable it is thought to be. Through this, inductivism is also connected with the views of methodological instrumentalists.

It seems to us reasonable to restrict the use of the term 'inductivism' to doctrines of this kind. We do not see any point in labelling as 'inductivist' all those views which allow of a role in science for induction – we apply it only to those views which exaggerate, simplify, and misconceive this role. Similarly, we do not denote all those views which allow of a place in science for deduction as 'deductivist'.

1.2. What could genuine deductivist doctrines be like? We could try to give definitions of deductivism which parallel our definitions of inductivism. Thus, *deductivism₁* can be defined as the purely rationalistic doctrine according to which scientific knowledge is based only on a kind of intellectual intuition and deduction. *Deductivism₂*, which was rejected already by Aristotle, claims that scientific laws and theories are deduced from observations. Finally, *deductivism*$_3$ asserts that the testing of scientific hypotheses is a purely deductive process which does not involve any kind of induction.

Popper's basic intention has been that of developing a deductivist$_3$ theory of the methodology of science (cf. Popper, 1959, 1963, and 1972). The main target of his 'anti-inductivist' attack has been the doctrines that we called inductivism$_2$ and inductivism$_5$. Popper basically accepts the method of hypotheses: in science, he argues, bold conjectures or highly falsifiable theories are tested by deducing from them the severest possible test statements. A theory which entails a false test statement is refuted, by *modus tollens*. A theory which passes severe tests is *corroborated*, but highly corroborated theories cannot be regarded as proved to be true or probable (cf. Popper, 1959, p. 33). However, if the appropriate methodological rules are followed, "we can have strong and reasonably good arguments for claiming that we may have made progress towards the truth" (cf. Popper, 1972, pp. 57–58).

It has been claimed that Popper's notion of corroboration reveals that his theory of science cannot be properly characterized as deductivist$_3$. Thus, Salmon has argued that "*modus tollens* without corroboration is empty; *modus tollens* with corroboration is induction" (cf. Salmon, 1966, p. 26; 1968, p. 28). In other words, if degrees of corroboration justify a principle of selection of unfalsified hypotheses, then "corroboration is a nondemonstrative form of inference" (Salmon, 1966, p. 26). One possible counterargument to this is to say that the notion of corroboration can be defined by using only concepts of deductive logic (cf. Watkins, 1968, p. 65). However, this does not yet refute Salmon's argument, even if Stegmüller (1971) thinks so. To do this, a Popperian has to admit, as Watkins (1968) and Popper (1972), pp. 82–84, indeed do, that corroboration appraisals are analytic, and do not have any predictive implications. That is to say, to appraise a theory as highly corroborated does not involve any prediction of its continued success in the future. "It says nothing whatever about future performance, or about the 'reliability' of a theory" (cf. Popper, 1972, pp. 18–19). In short, corroboration does not indicate the *reliability* or *trustworthiness* of a theory (which is what Lakatos (1968) calls the acceptability$_3$ of a theory).

To see whether this line of thought provides an adequate answer to

Salmon's argument, let us consider it in greater detail. It seems that when Salmon (1966) speaks of a method for *selecting* hypotheses, he primarily has in mind the *cognitivist* idea of the tentative and provisional acceptance of hypotheses for prediction and explanation (cf. Chapter 9.3). According to this view, the tentatively accepted hypotheses are still fallible, but we may have good 'inductive' grounds for the inclusion of them into the rational corpus of scientific knowledge. It is clear that Popper's ideas stand in sharp contrast with the cognitivist analysis of scientific inference. In his view, theories are never accepted as true – not even tentatively or provisionally. They are only accepted for the purposes of testing, or for the subjection to further criticism (cf. Popper, 1959, p. 419). However, the full force of Salmon's argument is that *any* method for selecting hypotheses is bound to be nondemonstrative, and, consequently, no theory of scientific inquiry which includes such a method cannot be regarded as deductivist₃. Moreover, even if we admit that scientific theories or hypotheses are never tentatively accepted as true, they are still accepted for practical purposes. It thus seems that Popperian deductivism₃ has to leave unanswered a central problem in the rational reconstruction of scientific practice, viz., the problem of the rationality in the trust in scientific theories that are applied in technology (cf. Lakatos, 1968).

In his latest work, Popper has tried to refute this objection (cf. Popper, 1972, pp. 18–29, 82–84). He defends corroboration as a method for selecting hypotheses for practical action: if the degree of corroboration of hypothesis h_1 in the light of the critical discussion at time t is greater than that of h_2, "we shall base our theoretical predictions as well as the practical decisions which make use of them upon h_2 rather than h_1." However, Popper argues, from a rational point of view, we should not *rely* on any theory for practical action, for "no theory has been shown to be true, or can be shown to be true." A theory cannot be *reliable*: there can be no "good reasons" for expecting that "it will in practice be a successful choice." Reliance on theories

would entitle us to rely on science; whereas today's science tells us that only under very special and improbable conditions can situations arise in which regularities, or instances of regularities, can be observed.

Therefore, Popper contends, there is no *absolute reliance*, "but since we *have* to choose, it will be 'rational' to choose the best-tested theory." In other words, we should *rationally prefer* as basis for action the best-tested theory, that is, the theory with the highest degree of corroboration.

Popper's position can be summarized by saying that he denies the legitimacy of both what cognitivists call *inductive inference* and what behaviouralists call *inductive behaviour*. He admits that hypotheses or theories *are* selected for practical action, but claims that the preference for one hypothesis over another does not involve any inductive element. Consequently, Popper's notion of the degree of corroboration is not a measure of *inductive* support, and his theory of scientific inquiry is characterizable as deductivist$_3$.

Popper's answer to Salmon and Lakatos can thus be taken to show that his theory of corroboration has nothing to do with induction. At the same time, it shows how implausible a full-blown deductivist$_3$ theory of scientific inquiry is. Even if cognitivism were rejected, reliance on science in technological applications is still an every-day fact. Of course, this reliance is not 'absolute' in any sense; still, it may have good reasons. The rationality of action or of practical decisions depends ultimately upon the estimated chances of attaining the objectives of action within the limits of available resources. Thus, if the consequences of a decision are open to a serious risk, it need not be rational to base one's decision upon the best-tested theory. Consequently, Popper's discussion of practical action is considerably oversimplified. At the same time, it makes a complete mystery of the trust and deliberate reliance on science.

The fundamental problem which Popper fails to answer is this: why is it *rational* to base one's practical decision upon the best-tested theory, if there are *no good reasons* for expecting that it will be a successful choice? One answer might be that the best-tested theory is in fact the one closest to the truth, so that no better choice is available. However, as Popper (1972), p. 103, admits,

degree of corroboration is at best only an *indicator* of verisimilitude, indicator of how its verisimilitude *appears* at the time *t*, compared with another theory.

Consequently, degrees of corroboration do not give any objective justification for Popper's principle of rationality of practical action. This principle is thus an unjustified methodological rule. This fact has serious consequences for the evaluation of Popper's theory of scientific inquiry. As N. Maxwell (1972) has pointed out, in failing to justify his own methodological rules, Popper fails also in explaining why we should value 'scientific' theories higher than others, that is, in answering adequately his fundamental 'problem of demarcation'.

1.3.   It seems that no deductivist$_3$ theory of science can give a complete account of those aspects of science which involve the acceptance or appraisal of the trustworthiness of theories on the basis of inconclusive evidence. This does not imply that such an account has to be inductivist in any of the senses defined above. "Scientific research seems to follow a *via media* between the extremes of inductivism and deductivism," as Bunge (1963), p. 151, has contended. Induction has a role in the validation of scientific knowledge, but even this role (which is not its only one) is much more complex than the inductivists have assumed.

Our approach to induction in this book is non-inductivist in all the five senses defined above. Against inductivism$_i$, $i=1, 2$, we have emphasized the role of hypothetical theories in science in general, and in the reinterpretation of observational statements. We have studied the inductive effects of the introduction of theoretical concepts, and thus connected the problems of induction and conceptual change. Against inductivism$_i$, $i=3, 4, 5$, we have emphasized the importance of accounting for the *theoretical support* of laws in addition to their observational support. Thus, we have claimed that scientific laws receive inductive support not only from 'below', i.e., from lower-level generalizations or positive instances, but also from 'above', i.e., through their connections with higher-level theories or nomological networks. Against inductivism$_5$, in particular, we have argued that the possible measures of support are not probabilistic in the narrow sense, but rather non-additive functions of probabilities. These probabilities, in turn, are not determined on purely *a priori* grounds, but depend, through two extra-logical parameters, on presuppositions of induction.

It should be mentioned at this stage that our emphasis on what we have called *hypothetico-inductive inference* does not involve a commitment to any form of inductivism. This inference is not an inference from observations *to* hypothetical theories, but, on the contrary, an inference which *starts from* hypothetical theories. Thus, the idea of hypothetico-inductive inference stands in contrast to both inductivism and deductivism. This point is discussed in greater detail in the next section.

2. HYPOTHETICO-DEDUCTIVE AND HYPOTHETICO-INDUCTIVE INFERENCE

In earlier chapters, we have several times alluded to the difference between hypothetico-deductive and hypothetico-inductive inference. A simple characterization of these two kinds of inferences is as follows: they represent deductive and inductive inferences, respectively, from hypothetical premises. In this section, we try to make the difference between these types of inferences sharper, and to indicate what kinds of methodological situations they may cover. We also discuss the questions of the relations of these inferences to deductivist and inductivist doctrines, as well as to the notions of deductive and inductive systematization. As we shall see, these relations are not as simple and clear-cut as one might expect.

2.1.   According to the Aristotelian ideal of science, scientific theories are axiomatic or deductively closed systems of general statements. This view naturally suggests that induction, or non-deductive inference in general, can in science play only the role of an inference which proceeds from singular observational statements to theories (and, possibly, directly to other singular statements). Consequently, as we saw in the preceding section, induction has sometimes been conceived of as a method of scientific discovery leading from observational evidence to laws and theories. When the untenability of this inductivist$_2$ doctrine became clear, it was only natural to suggest that induction is a method of validation of scientific theories on the basis of observational evidence. In other words, even if the discovery of theories is a non-inductive process, hypothetical theories are tested and justified inductively. It also seems tempting to suppose that deduction plays an indispensable role in the testing of theories by observational data. Deduction leads us from general theories to lower-level generalizations and to singular observational test statements. On the other hand, induction is an inverse process to deduction in the straightforward sense that it leads us, though non-demonstratively, from observational evidence to laws and theories. This non-inductivist$_2$ account of deductive theories can be briefly summarized as follows: Intratheoretical relations between statements forming a scientific theory are deductive, while the relation between a theory and its evidential basis is deductive in one direction and inductive in the converse direction.

The view which we just have sketched out is the basic idea in the *hypothetico-deductive method* of testing scientific hypotheses, which was ably discussed by such 19th century theorists as Whewell and Peirce. In this view, scientific inquiry follows the *method of hypothesis*: A hypothesis *h* is first introduced or invented to account for or to explain some initial observational data e_0. This hypothesis is then tested by deducing a new observational statement *e* from it. If *e* proves to be false, *h* is falsified. If *e* proves to be true, *h* is confirmed or supported by *e*. For example, hypotheses which are general statements are supported by their true substitution instances. Highly supported hypotheses, i.e., those which entail sufficiently many true test statements, can be tentatively accepted.

It is obvious that the above characterization of the hypothetico-deductive method needs a great deal of further elaboration. One way of generalizing it is provided by the notion of *deductive systematization* which we defined in Chapter 1.2. We stated there that a theory *T* establishes deductive systematization between two observational statements *e* and *h* if and only if *h* is deducible from '*e* & *T*' but not from *e* alone. A wide variety of important methodological situations can be characterized in terms of this notion. Thus, *h* may be an observational law or a singular statement which is deductively explained by the theory *T* and the initial conditions expressed by *e*. The statement *h* may also be a prediction which can be deduced from *T*, assuming *e*. Both of these cases represent *hypothetico-deductive inference* in the sense that *h* is deduced from the hypothetical theory *T*, assuming *e*. One can also suggest that the support that a theory receives from observations should be somehow dependent upon its capacity to establish deductive systematization between observational statements. In fact, essentially this was already Whewell's (1847) view of the matter; besides prediction, he mentions the 'consilience of inductions', that is, the establishment of deductive systematization between generalizations, as a test of a theory.

It seems fair to say that the idea of hypothetico-deductive inference contains an important insight into many central aspects of scientific inference. Nonetheless, the notion of inductive support that is more or less implicit within it seems to be hopelessly simplified, and the attempts to clarify this notion have not been very successful. Some philosophers have tried to develop a deductivist logic of science which is based on the notion of deductive systematization without any recourse to induction,[5] while some others have concluded that it is not possible, on the basis of the hypothetico-deductive account, to develop a satisfactory characterization of the role of theories within inductive scientific inference. Some of these problems are discussed in the next two sections. Here, we illustrate the limitations of the hypothetico-deductive account by contrasting it with the notions of inductive systematization and hypothetico-inductive inference.

2.2. In *Posterior Analytics*, I, 2, Aristotle contended that scientific knowledge is demonstrative and must proceed from premises that are, among other things, true and better known than the conclusion. However, he was well aware that, apart from 'scientific syllogisms' which satisfy such requirements, there also exist other kinds of syllogism. The hypothetico-deductive view of scientific inference is often seen as an important step in the liberation from the Aristotelian ideal of demonstrative science. In this respect, it is contended that one cannot have demonstrative but only probable knowledge in empirical sciences. It is also recognized that all the deductive inferences in science need not satisfy Aristotle's requirements for a scientific syllogism. Thus, according to this view, a central role within scientific inference is played by hypothetico-deductive inference, in which the premises are neither known to be true nor necessarily better known than the conclusion.

Not all philosophers have found the idea of hypothetico-deductive inference to be appealing. John Stuart Mill made the following charge against syllogistic reasoning: "No reasoning from generals to particulars can, as such, prove anything: since from a general principle we cannot infer any particulars, but those which the principle itself assumes as known" (cf. Mill, 1843, Book II, Ch. III, 2). In other words, the universal premise of a syllogism cannot be known to be true unless the conclusion is already known to be true. Therefore, Mill concludes, the real premises, say of a syllogism proving the mortality of the Duke of Wellington, are the particulars asserting the "mortality of John, Thomas, and others." "Not one iota is added to the proof by interpolating a general proposition" asserting that all men are mortal (*ibid.*, Book II, Ch. III, 3). In Mill's view, all inference is from particulars to particulars. He is ready to admit that such an inference may pass through a generalization as an intermediate stage, but then the real *inference* resides in the ascent to the generalization. The remaining deduction from the generalization to the particular conclusion is not an inference but merely a "deciphering our own notes" (*ibid.*, Book II, Ch. III, 3).

Mill's theory of syllogism is based on the mistaken idea that science seeks and can attain demonstrative knowledge. He also assumes that all inference must proceed from premises which are known to be true. This leads him to reduce deduction to induction. Consequently, his remarks on the 'hypothetical method', in which hypotheses are 'verified' by deducing 'known truths' from them (*ibid.*, Book III, Ch. XII, 4) cannot be taken too literally; according to his account of reasoning, the real inference can not proceed deductively from the hypothesis itself, but rather inductively from some particulars known to be true.

Mill is a representative of the traditional doctrine that inference should always lead us from truths which are 'directly' known, to other truths which consequently are known only 'indirectly'. The idea of hypothetico-deductive inference liberated deduction from such a role. Still, the Humean view of induction as an inference from known to unknown, or from certain knowledge to uncertain, was preserved. It seems that this doctrine has prejudiced the role which inductive inference is allowed to play within the hypothetico-deductive account. When combined with an empiricist theory of knowledge, it suggests that induction should always proceed from observational knowledge *to* hypothetical theories or to predictions concerning future events.

However, if deduction can proceed from hypothetical premises, why could not induction do so as well? There is no need to assume that the premises of an inductive inference are true or known to be true – no matter how this inference is explicated. This is, indeed, the position we have adopted in this book. We do not claim only that induction may proceed from fallible observational statements, but from conjectural theories as well. Thus, we have said that the situations in which a hypothetical theory occurs among the premises of an inductive argument represent *hypothetico-inductive inference*. This notion is, therefore, an obvious generalization of the hypothetico-deductive inference. Deduction is simply replaced by induction, and it is recognized that hypotheses – and theories in particular – may have both inductive and deductive connections with other statements. In a similar way, we have generalized the notion of deductive systematization by saying that a theory T establishes *inductive systematization* between two observational statements e and h if and only if h is inducible from e and T, or from e in the light of T, but not from e alone (cf. Chapter 1.2).

The notion of inductive systematization covers a wide variety of methodological situations. For example, T may be a theory which explains h inductively, given e. Here h may, for example, be an observational generalization or a singular statement. We have studied, in Chapter 7, situations in which h is a generalization and induction is explicated by the notion of positive relevance. Furthermore, explanations of singular statements by probabilistic theories belong to inductive systematization in our sense.[6] Another kind of situation is the case in which h is prediction which is inductively based on T and e. Here h may again be singular or universal, and T may be a deterministic or probabilistic theory.

In general, we have studied situations in which we can evaluate the 'inductive effect' of introducing a theory employing new concepts with respect to observational generalizations. This effect is measurable in various ways, of which we have studied inductive probabilities, transmitted information, systematic power and degrees of support or of corroboration. Among other things, this approach gives us a method of measuring amount of theoretical support which laws can obtain from higher-level theories.

In Chapter 8, we also mentioned the possibility of discussing some aspects of the inverse problem of measuring the support which theories can obtain from observations. Even if we have not systematically studied this problem in this work, it is obvious that our approach can, in principle, be extended so as to cover it. (See also the next section.) For example, if a theory T is found to be positively relevant to an observational statement h, relative to e, then h is positively relevant to T, relative to e. Thus, one can develop a hypothetico-inductive method of testing hypotheses which is based, for example, upon some symmetrical explicate of inductive inference (such as positive relevance). Such a theory has, of course, an important difference from the hypothetico-deductive notion of support: even theories lacking deductive observational consequences (i.e., theories which do not establish deductive systematization with respect to observational statements) can receive observational support. It also follows that such theories can establish inductive systematization between observational statements.[7]

These remarks already suffice to indicate some of the central differences between the notions of hypothetico-deductive and hypothetico-inductive inference, as well as of deductive and inductive systematization. Still another way of contrasting hypothetico-inductive inference with traditional views is as follows. In the Aristotelian view, scientific theories are conceived of as deductive systems, that is, as networks of general statements connected by a relation of deducibility. In the cases of hypothetico-inductive inference which we have discussed, a hypothetical theory occurs as a premise and a universal law as the conclusion. Accordingly, it can be said that here the theory and the law concerned belong to a network of statements between which there need not be only deductive, but also inductive relationships.

It is obvious that the claim of the existence and the methodological importance of hypothetico-inductive inference is incompatible with deductivist doctrines. It is perhaps less obvious that it can be interpreted as antithetical to inductivist doctrines as well. Some critics might even suspect that this claim commits us to a kind of disguised inductivism. However, this suspicion is not warranted, since hypothetico-inductive inference need not involve any of the characteristic features or assumptions of the inductivist doctrines. Thus, this inference does not start from observations but rather from theories. It is not a method by which theories or laws are discovered. Instead, it provides a method for evaluation of the observational *and* theoretical support of general statements. However, even if the notion of support or corroboration is legitimate as such, its measures are not probabilistic in the narrow sense of inductivism$_5$. It is not necessary, moreover, to take any definite standpoint in answering the question of the origins of the hypothetical theories occurring as premises of hypothetico-inductive inference. They may be conjectures which have more desirable features than their rivals – for example, higher content, greater explanatory power, or higher degree of corroboration on the basis of higher-level theories. They may be independently testable, deductively or inductively. They might also be theoretical presuppositions only, which are considered as unproblematic and tentatively accepted in some situation. Or, finally, they need not necessarily contain factual statements at all, but only statements expressing the conceptual or definitory connections between the old and new vocabularies. In short, even the provisional acceptance of the premises of a hypothetico-inductive argument need not mean committing oneself to any form of inductivism.

2.3.   Hypothetico-inductive inference and inductive systematization are made especially interesting by the justification which they give to the use of theories and theoretical concepts. As we have seen in Chapter 9, theoretical concepts can be logically indispensable for inductive, but not for deductive systematization of observational statements. The ingenious insight of Hempel (1958) saw that this might be the case. It was also Hempel who was the first to introduce the general notion of inductive systematization – even if some germs of this idea are discoverable in Whewell's notion of the consilience of inductions. Nevertheless, as we saw in Chapter 1, Hempel's account of inductive systematization is not satisfactory, since it raises a transitivity dilemma which cannot be solved within it. Consequently, it is in order to repeat there the basic idea of our solution of this dilemma, and to contrast our account of inductive systematization with that of Hempel's.

In his early study in confirmation, Hempel advanced the thesis that as a rule theories using theoretical terms do not yield observational predictions from finite observation reports without 'quasi-inductive' steps of reasoning (cf. Hempel, 1945, pp. 27–30). He also remarked that this may hold not only for theories, but for some observational generalizations too.[8] In other words, he claimed that, in connection with prediction, theories do not establish deductive but inductive systematization. Therefore, he thought, hypothetico-deductive inference is only a simplified account of the inference which, in fact, leads to predictions by a series of deductive and inductive steps.[9]

It is illustrative to compare this view of Hempel's with another criticism of scientific deduction, viz. John Stuart Mill's theory of syllogism. Both have contended that the seemingly deductive inference from a theory T and a report of past observations e to an observational prediction h is at bottom inductive. Mill thought that here h is inducible from e alone. Moreover, h is so inducible if and only if the generalization T is inducible from e (cf. Mill, 1843, Book II, Ch. III, 5). Accordingly, T is only a register of inductions already made, and a formula for making more (*ibid.*, Book III, Ch. III, 4). It follows that the introduction of T is of no help here in the sense that T is not indispensable for the inference from e to h. Consequently, Mill's view leads to the rejection of the idea that theories or generalizations could be indispensable for inductive systematization. Hempel's conclusion is diametrically opposed to this; he thinks that *h* may be derivable from *e* and *T* by deductive and inductive steps of reasoning, and that *T* may be indispensable for this inference (cf. Hempel, 1958).

One can agree with Hempel that not all scientific predictions are obtained deductively from theories and initial conditions, but we nevertheless find his emphasis on inductive systematization somewhat overstressed here. At least if one operates with the standard conception of theories as sets of statements containing both purely theoretical statements and correspondence rules, then theories usually, though not always, have deductive observational consequences and yield deductive observational predictions. For this reason, the notions of hypothetico-deductive inference and deductive systematization are legitimate, even if they are not comprehensive enough to cover all methodological situations.

As was seen in Chapter 1.1., Hempel took inductive systematization to be established *indirectly* by an argument which contains at least *two steps*. These steps proceed through theories, so to say, and *via* theoretical concepts. However, this idea is questionable, since inductive inference does not seem to have any simple or modified transitivity properties.[10] In particular, Hempel's examples of inductive systematization were reasonable only on the assumption that the conditions of Converse Entailment and Special Consequence can hold at the same time. However, it is known that this cannot be the case for any reasonable explicate of inductive inference. Our solution of this transitivity dilemma can be put simply as follows. We reduce Hempel's indirect two-step arguments to a *one-step* inductive argument, which proceeds directly to the conclusion from the theory and the initial conditions, or from initial conditions in the light of the theory. At the same time, we account for the conceptual change from observational to theoretical language. In brief, our notion of hypothetico-inductive inference is the essential basis of our solution. The dilemma disappears when we replace Hempel's idea of reasoning as a chain of deductive and inductive steps by this notion.

The transitivity dilemma can be seen to appear also in other connections, and it remains to be studied whether our solution of it, in a particular case, throws some light on the other cases. We do this in Section 4 below.

3. THE ATHEORETICAL THESIS

3.1.   As we have seen above, no theory of induction should be expected to provide a logic of scientific discovery. The discovery of scientific laws and theories is not a purely inductive process. However, when it was realized that inductivism$_2$ is a mistaken doctrine, the problem of accounting for the support or trustworthiness of theories, laws, and their predictions still remained for the theorists of induction to solve. A satisfactory solution of this problem should be able to account for the relation between general knowledge and singular observations. Since scientific inquiry is neither purely deductive nor purely inductive, it should also be able adequately to treat those aspects of scientific inference in which is involved an interplay of deduction and induction.

Such a solution, if it can be formulated, might be called an *inductive logic of theories*. Below, we discuss some partial solutions to this problem; they serve to illustrate the obstacles to be faced in any attempt to develop an inductive logic of theories. At the same time, we try to suggest how at least some of these obstacles can be overcome.

One way of approaching these problems is to try to define a measure C which adequately expresses the degree of 'confirmation', 'support', or 'corroboration' of theories and laws (cf. Chapter 8). The 'Cambridge school' endeavoured to accomplish this by developing a probabilistic theory of induction in which C is identified with probability P (cf. our discussion of inductivism$_5$ in Section 1).

In his discussion of the development of inductive logic, Lakatos has given the name '*the Jeffreys-Keynes postulate*' to the assumption that the prior probability $P(g)$ of any consistent theory or law g is non-zero (cf. Lakatos, 1968, pp. 330–334). Together with the assumption that $C = P$ this postulate implies that theories can receive non-zero degrees of support on the basis of finite observational evidence. However, in the system of inductive logic which Carnap developed in the early forties, it proved that $P(g) = 0$ for any genuinely universal statement (cf. Carnap, 1950, pp. 570–571). According to Lakatos, Carnap had then the following possibilities of solving the inconsistency between this result and the Jeffreys-Keynes postulate:

(1)   To accept $C(g) > 0$, and either (a) to reject $P(g) = 0$, or (b) to reject $C = P$.

(2)   To accept $C(g) = 0$.

(3)   To restrict the domain of C to singular statements.

Lakatos notes that "Carnap tried possibilities (1a), (2), and (3), *but never* (1b)," while "Popper, on the other hand, thought that (1a), (2), and (3) had to be ruled out; (1b) *was the only solution*" (cf. Lakatos, 1968, p. 332). Popper's solution (1b) was his non-probabilistic measure of corroboration (cf. Chapter 8.2). Carnap's theory of 'qualitative instance confirmation', that is, his attempt to evaluate the reliability of laws by the probability of their unobserved instances, was a solution along the lines of (2) (cf. Carnap, 1950). Subsequently, he tried also to develop a system of inductive logic along the lines of (1a), i.e., so that universal laws can receive non-zero probabilities (cf. Carnap, 1963, p. 977). However, his basic intuition was that $P(g) = 0$ is not only acceptable but also desirable. He came to think that inductive logic should be a guide in determining practical decisions, and that inductive probability is a logical concept corresponding to the quasi-psychological concept of 'rational degree of belief' (cf. Carnap, 1962a and 1971a). In practical decision situations, there is no longer a need for a probability of laws or theories: inductive logic is essentially concerned with betting on particular predictions. Solution (3) thus seems to be viable. This idea, along with the contention that the rational betting quotients for particular predictions are independent of the available scientific theories, has been termed the *weak atheoretical thesis* by Lakatos (cf. Lakatos, 1968, pp. 335–359, 361–375). This thesis solves the problem connected with the Jeffreys-Keynes postulate my making theories and laws dispensable in inductive logic.

It seems to us that all these attempts of Carnap and Popper are unsatisfactory. We have already criticized Popper's solution (1b) in Chapter 8 (cf. also Section 1). The fact that Carnap's theory of qualitative instance confirmation is not satisfactory has convincingly been shown, *inter alia*, by Hintikka (1965a).[11] In our view, the basic mistake common to both Carnap and Popper is that they reject the possibility that $P(g) > 0$ for universal statements g.[12] $P(g) = 0$ implies that $P(g/e) = 0$ for any evidence e, so that one cannot discuss and compare the support of general statements in probabilistic terms. Thus, Carnap's (1962a) definition of support in terms of positive inductive relevance trivializes in the case of laws and theories. (Note, however, that this suggestion is a solution where $C \neq P$, which Lakatos (1968), p. 332, wrongly claims to have been inconceivable for Carnap.) Similarly, Popper's measures of the degree of corroboration of theories also trivialize – unless *ad hoc* additions concerning the 'fine structure' of theories are made.[13] Therefore, it can be said that the assumption that $P(g)=0$, for universal statements g, has been at least one decisive factor leading Carnap to reject an inductive logic of *theories*, and Popper to reject an *inductive* logic of theories.

To express this in different terms, we may say that Carnap's inductive logic fails to appreciate the role of theories and laws within scientific inquiry, while Popper's deductivism fails to account for the rationality of the inductive acceptance of theories for the purpose of making predictions (cf. also Section 1).

An approach of a new kind is thus needed.

3.2.   Let us consider the atheoretical thesis in more detail. As Lakatos (1968), pp. 340–343, emphasizes, Carnap never accepted this thesis in the strong form that claims theories to be dispensable both in the logic of discovery and in the theory of induction. In Carnap (1950), he only thought that the judging of discovered laws can be reduced to the judging of their singular predictions. Accordingly, he thought that singular predictive inference is more important than inductive generalization "not only from the point of view of practical decisions but also from that of theoretical science" (cf. Carnap, 1950, p. 208). For Carnap, this inductive inference means the same as the assigning of probability values to singular predictions. He concludes that

We see that the use of laws is not indispensable for making predictions. Nevertheless it is expedient, of course, to state universal laws in books on physics, biology, psychology, etc. Although these laws stated by scientists do not have a high degree of confirmation, they have a high qualified-instance confirmation and thus serve as efficient instruments for finding those highly confirmed singular predictions which are needed in practical life. (Carnap, 1950, p. 574.)

John Stuart Mill had already claimed that "all inference is from particulars to particulars." According to him, singular predictions are not inferred *from* universal laws, but from particular facts. The universal law, which itself is established by inductive generalization, serves as a *rule of inference* (or 'inference ticket', to use Ryle's term) in accordance with which this inference is drawn (cf. Mill, 1843, Book II, Ch. III, 4).

Both Mill and Carnap here advocate an instrumentalist view of universal laws as 'efficient instruments' for finding singular predictions. This view is, of course, unacceptable to scientific realists, or to cognitivists who would like to incorporate universal laws into their theory of the world. For them, scientific laws are not only 'instruments' for attaining predictions of particular facts; rather, these laws themselves represent the basic part of our knowledge of the world. This is so, because, for a realist, the universal laws which science seeks to establish express objective regularities or uniformities which exist 'out there' in the world. It is also precisely for this reason that laws can have genuine explanatory power. The instrumentalist view of laws can be plausible at all only if one assumes, following Carnap, that the world is 'atomistic', i.e., that the world consists of individuals between which no general laws hold good. As we saw above, however, a scientific realist would include objective regularities among the 'inhabitants' of the world.

Hintikka and Hilpinen have pointed out an interesting connection between universal generalization and singular predictive inference.[14] They have shown that, in every theory of inductive acceptance which satisfies the conditions of consistency and logical closure (of the set of acceptable hypotheses) and which requires high probability as a necessary condition of acceptance, the following holds:

(AS)   A singular hypothesis $A(a_i)$ is acceptable on evidence e iff the generalization $(x)A(x)$ is acceptable on e.

More generally, $A(a_i)$ is acceptable on e if and only if $A(a_i)$ is implied by the conjunction of e and acceptable generalizations. Theories of acceptance satisfying (AS) can be used to justify a standpoint which is in sharp contrast with Mill's and Carnap's views. Even though Mill thinks, in accordance with (AS), that "the experience which justifies a single prediction must be such as will suffice to bear out a general theorem" (cf. Mill, 1843, Book II, Ch. III, 5), he accepts generalizations only as "an intermediate halting-place for the mind, interposed by an artifice of language between the real premises and the conclusion" (*ibid.*, Book II, Ch. III, 6). Moreover, principle (AS) can be used for arguing against Carnap that the judging of singular predictions can and should be reduced to the judging of universal laws. (This is, indeed, one of the reasons for our focusing on generalizations rather than on singular statements within our system.)

The atheoretical thesis does not claim only that it is sufficient, for the purposes of inductive logic, to compute the values $P(h/e)$ for singular statements h. Another aspect of it is the assumption that evidence e is likewise singular. In other words, it is assumed that the inductive probabilities of singular predictions are independent of the available scientific theories. This presupposes that the evidence for predictions "can in general be identified and established without even the tacit acceptance and use of laws," as Nagel (1963), pp. 802–3, points out. Nagel (1963), p. 805, also argues that

it is not the *instantial evidence* for the laws of mechanics, but rather the *laws of mechanics* themselves, which must be included in the evidence for the hypothesis of another sunrise.

A similar suggestion has been made by Putnam (1963). We have also argued the importance of using theories as starting-points of hypothetico-inductive inferences (cf. Section 2). Moreover, we argued that even singular evidence statements can be theory-laden, at least in the sense that they are reinterpreted in the light of the theoretical vocabulary. In short, if observational evidence is theory-laden, or if theories or theoretical background assumptions are used as evidence of inductive arguments, the use of general laws in inductive logic cannot be eliminated, and consequently the atheoretical thesis collapses.[15]

In his interesting reply to Putnam's and Nagel's arguments for using theoretical laws as evidence, Carnap remarks that he has developed inductive logic only for simple qualitative observational languages, and admits that in a theoretical language inductive probabilities can be influenced by "the primitive magnitudes of the theoretical language and the postulates and correspondence rules" (cf. Carnap, 1963, pp. 986–989). This answer implies, in effect, that Carnap is ready to admit that in a system of inductive logic for the total (i.e., observational and theoretical) language of science – if only such a system were developed – neither the weak atheoretical thesis nor the criterion of linguistic invariance (cf. Chapter 10) holds true. This also shows that much of Lakatos's criticism of Carnap as a defender of the weak atheoretical thesis does not apply to Carnap's (1963) views.

3.3. In this work, we have tried to develop an inductive logic for simple theories containing qualitative monadic predicates. Hintikka's system of inductive logic, which we have applied here, provides us with probability measures P which give non-zero probabilities for genuinely universal statements. It is characteristic of our approach that, among Lakatos's alternatives (1)–(3) above, we have made a choice which was rejected both by Carnap and Popper – and is left unmentioned by Lakatos. This choice is to accept both (1a) *and* (1b) at the same time. In other words, in our system universal statements receive non-zero probabilities, but the degree of support or corroboration C is not identified with posterior probability. Nonetheless, this is not enough. To develop an inductive logic of theories, one has to take one further step. The probability measure P must be defined for the language $L(\lambda \cup \mu)$ containing both observational and theoretical predicates. In particular, it should be possible to give probability values to observational generalizations on the basis of observational and theoretical evidence, as well as to hypotheses containing theoretical concepts. This is, indeed, what is accomplished by our technical development in Chapters 2–5.

Several authors have made the remark that the development of an inductive logic for languages containing theoretical concepts would be desirable.[16] But it has also been claimed that this task is impossible to accomplish. Thus, Barker has argued that

a transcendent hypothesis of the second kind, one containing theoretical predicates, clearly cannot be confirmed by induction, whether that induction be enumerative or eliminative. (Barker, 1957, p. 97.)

By "a transcendent hypothesis of the second kind" Barker means, in effect, a statement containing non-evidential theoretical concepts. One might accordingly interpret Barker's thesis as claiming that one cannot compute degrees of confirmation or support of hypotheses g in a language $L = L(\lambda \cup \{P\})$, where '$P$' is a non-evidential predicate, on the basis of observational evidence e in $L(\lambda)$. Although we have not given any formula for calculating such values, one can easily see how the probability of g in L on evidence e in $L(\lambda)$ can be determined within our system. To see this, let us suppose that

$$\vdash g \equiv C_1^r \vee \dots \vee C_k^r,$$

where C_i^r, $i = 1, \dots, k$, are some constituents of L. Then

$$P(g/e) = \sum_{j=1}^{k} P(C_j^r/e),$$

where

$$P(C_j^r/e) = \frac{P(C_j^r)\, P(e/C_j^r)}{P(e)}.$$

In Chapter 3, we saw how the products

$$P(C_j^r)\, P(e/C_j^r)$$

can be computed in the case of non-evidential theoretical predicates. Moreover, here

$$P(e) = \sum_j P(C_j^r)\, P(e/C_j^r),$$

where the sum is taken over those constituents C_j^r of L that are compatible with evidence e. (If the Ct-predicate Ct_h of $L(\lambda)$ is exemplified in e, such a constituent C_j^r cannot rule out both Ct_h & P and Ct_h & $\sim P$.) Thus, we can compute $P(e)$ and, hence, also $P(g/e)$ whenever g is in L and e is in $L(\lambda)$.

We can conclude that posterior probabilities of hypotheses containing theoretical concepts can be computed within our system. These probabilities can then be used in evaluating the degree of support or corroboration of these hypotheses. We do not systematically study these problems here. It suffices to remark that cases where e in $L(\lambda)$ supports g in L e.g. in the positive relevance sense can easily be found.[17] This conclusion refutes Barker's claim.[18]

4. CONVERSE DEDUCTION AND INDIRECT SUPPORT

It was seen above that it is possible to develop an inductive logic of theories which is capable of dealing with general theories containing theoretical concepts. Such a logic provides us with inductive probability measures, depending on two extra-logical parameters, which can be used for the definition of different measures of support for theories and generalizations. In particular, we have shown how generalizations can receive support from observations and from theories, and how various situations belonging to hypothetico-inductive inference and inductive systematization can be handled adequately.

We have seen that the transitivity dilemma arising from Hempel's account of inductive systematization can be solved within our system, without recourse to any suspectible transitivity principle of inductive inference. Still other problems arise from the interplay of deduction and induction within scientific inference. These problems include the hypothetico-deductive account of the support of theories by observational tests, and the more complicated situations involving support which is 'indirect'. A fully-developed inductive logic of theories should to be able deal with these problems, but, at the moment, there exists no satisfactory general solution to them. What is more, Hesse has concluded that "the problem of theoretical inference or confirmation is insoluble in terms of a deductive model of theories" (cf. Hesse, 1970b, p. 164). In this section, we try to discover what light our approach may shed on this question, and the extent to which Hesse's claim is justified.

4.1.   The 19th century champions of the method of hypothesis, such as Whewell and Jevons, emphasized that induction, viewed as a process of justification or testing of conjectural hypotheses, is a kind of *converse deduction*. In this view, hypotheses are tested by deducing observational statements from them, and by verifying these statements. If a hypothesis entails a true test statement, it is 'confirmed' or 'supported' to some degree. Thus, true substitution instances of a generalization support that generalization, and true laws deduced from a theory support that theory.

This idea is justifiable in terms of positive relevance as follows. Whenever $P(h) > 0$ and $P(e) < 1$, it is sufficient for $P(h/e) > P(h)$ that h entails e. More generally, suppose that a statement e is deducible from the conjunction of h and b, where e is not deducible from b alone. Then $h \vdash b \supset e$ and not $\vdash b \supset e$. If now $P(h) > 0$, statement '$b \supset e$' is positively relevant to h. In other words, if the support relation I is explicated in terms of positive relevance, then the following holds:

(CE)   *Converse Entailment*: If $h \vdash e$, then eIh

for all statements h and e such that $P(h) > 0$ and $P(e) < 1$. Moreover, CE implies that, for h, b, and e such that $P(h) > 0$,

(CE')   If $(h \& b) \vdash e$ and not $b \vdash e$, then $(b \supset e) Ih$.

While CE and CE' give only a sufficient condition of support in terms of deduction, it has sometimes been thought that confirmation-relations between statements can be *reduced* to converse entailment-relations. Thus, von Wright defines 'Confirmation-Theory' as the "theory of how the probability of a given proposition is affected by evidence in the form of propositions which are logical consequences of it" (cf. von Wright, 1957, pp. 117–118). It has also been thought that the qualitative concept of support or confirmation can be *defined* in terms of conditions corresponding to CE and CE'. Thus, referring to Ayer (1946), Hempel (1945), pp. 25–30, has defined the *Prediction Criterion of Confirmation* as follows:

Let H be a hypothesis, B an observation report, *i.e.* a [finite] class of observation sentences. Then (a) B is said to confirm H if B can be divided into two mutually exclusive subclasses B_1 and B_2 such that B_2 is not empty, and every sentence of B_2 can be logically deduced from B_1 in conjunction with H, but not from B_1 alone; (b) B is said to disconfirm H if H logically contradicts B; (c) B is said to be neutral with respect to H if it neither confirms nor disconfirms H.

Similarly, Popper (1959), pp. 266–267, suggests that

A theory is to be accorded a positive degree of corroboration if it is compatible with the accepted basic statements and if, in addition, a non-empty sub-class of these basic statements is derivable from the theory in conjunction with the other accepted basic statements.

He claims that this gives an adequate definition of 'positively corroborated' if it is required that the derived basic statements are "accepted as the results of sincere attempts to refute the theory." Moreover, if the restriction to *accepted* basic statements is omitted, this definition turns into a criterion of demarcation between 'scientific' and 'metaphysical' theories. According to this criterion, a theory is 'scientific' or 'empirical' if "the basic statement b_2 does not follow from b_1, but follows from b_1 in conjunction with the theory" (cf. Popper, 1959, note 1, p. 85). (Note that here the statement '$\sim b_2$ & b_1' is a potential falsifier of the theory in question.)

Stegmüller has recently given a definition of 'deductive support' which is based on Popper's ideas referred to above (cf. Stegmüller, 1971, pp. 31–35). In our terms, the essential content of this definition can be simply expressed as follows:

(DS)   Let e and h be statements in $L(\lambda)$, and let T be a theory in $L(\lambda \cup \mu)$. Then $\{e, h\}$ *supports* T if and only if
(a) $\{e, h\}$ is compatible with T
(b) T establishes deductive systematization between e and h
(c) h reports a severe test of T.

If no requirement is imposed upon the truth of e and h (or on their acceptance as true), DS gives only a definition of potential support. Note also that if condition (c) is omitted, DS is essentially the same as the above-mentioned Prediction Criterion of Confirmation.

One may say that a theory T is *testable* iff there are statements e and h in $L(\lambda)$ such that conditions (a) and (b) of DS are satisfied. (If (c) is satisfied too, we might say that T is severely testable.) Then

> T is testable iff T establishes deductive systematization with respect to $L(\lambda)$
>> iff T has non-tautological deductive consequences in $L(\lambda)$
>> iff T has potential falsifiers in $L(\lambda)$
>> iff T is 'directly verifiable' in the sense of Ayer (1946), p. 13.

4.2.   We now make some comments on the criticism which has been made against DS and, consequently, against the notion of testability based on DS.[19]

Stegmüller (1971) has contended that DS is inadequate in several respects, and that no one has as yet shown whether it can be modified to a satisfactory form. It seems to us, however, that of Stegmüller's five objections against DS only *one* has any force. This objection is the following: if $\{e, h\}$ supports T, then it supports *any* theory T' such that $T' \vdash T$. In other words, DS satisfies an unqualified form of the so called Converse Consequence Condition (cf. below). The second point of Stegmüller is based on an inadequate way of reading definition DS[20], the third has already been taken care of in our definition of DS[21], and the fourth is based on a logical mistake[22]. In his fifth point, Stegmüller claims that Goodman's paradox (cf. Chapter 10.5) arises also in connection with DS. In this paradox, it is shown that for any generalization h one can find another generalization h' such that h' is *equally well supported as* h on the same evidence e, but h and h' yield contradictory predictions. It is clear that this paradox can be formulated only in terms of a notion of support which is at least comparative. Moreover, there is nothing paradoxical in the fact that some evidence statement can support both h and h', or even both of two logically incompatible hypotheses. Consequently,

Goodman's paradox can not arise in connection with a qualitative notion of support such as DS.

Hempel (1945), along with Scheffler (1963), pp. 242–245, has rejected the Prediction Criterion of Confirmation (corresponding to clauses (a) and (b) of DS) as "too narrow to serve as a general definition of confirmation." Hempel claims that as a rule theories containing theoretical terms, do not yield observational predictions from past observations without 'quasi-inductive' steps of reasoning (cf. Section 2). He concludes that this criterion does not apply to theoretical hypotheses and not even to all observational sentences.

We can agree with Hempel that DS is too narrow as a *general* definition of support. However, as we remarked in Section 2, theories in $L (\lambda \cup \mu)$ *may* have consequences in $L (\lambda)$, and DS *is* applicable to such theories. Nevertheless, DS is too narrow for the reason that it takes into account only support on the basis of converse entailment relations.[23] One also needs measures of support that are applicable to the cases of inductive systematization. Indeed, our measures of support in Chapter 8 are of this kind. Moreover, it is very simple to justify such measures exactly in the same way as in the argument for Converse Entailment principle above. We noted above that e is positively relevant to h if e is deducible from h. However, e may be positively relevant to h even if e is *not* deducible from h. Since positive relevance is a symmetric relation, it is sufficient to assume that h is positively relevant to e. Definitions of support based on the notion of positive relevance are indeed much more general than the 'deductivist' notion of support defined by DS. Consequently, the notion of 'confirmability' or 'testability' corresponding to the positive relevance sense of support (i.e., T is 'confirmable' with respect to $L (\lambda)$ if there exists a statement e in $L(\lambda)$ such that T is positively relevant to e) is likewise more general than the 'deductivist' notion of testability based on DS. (Cf. also Chapters 8 and 9 above, and Niiniluoto, 1973b.)

This argument shows only that the theory of inductive support cannot be reduced to the study of converse entailment relations, so that DS cannot give a satisfactory general definition of support. It can be plausibly argued, however, that there does not exist any *general* definition of support applicable to all situations to be dealt with. Namely, one should not expect to find any unique explication for the notion of inductive support, since there are not only different intuitions about induction for a given inductive situation, but different inductive situations as well. These different situations may similarly require different notions of inductive support (cf., for example, Smokler, 1968). In view of this, one might suggest that DS or the Prediction Criterion serves at least as a first approximation of a restricted notion of inductive support applicable to the important situations in which theories or hypotheses are tested through their deductive observational consequences.

If we adopt this view on the status of DS, there is nothing fatal in Hempel's (1945) objection that the Prediction Criterion does not satisfy the following principle:

(SC)   *Special Consequence*: If eIh and $h \vdash b$, then eIb.

It cannot be required, as Hempel does, that any explication of the relation I satisfies Special Consequence SC, since we know that SC cannot be reasonably combined with the Converse Entailment principle (cf. Chapter 1.1). In particular, Special Consequence cannot be true in those situations where DS is applicable. This is reflected also in the fact that the positive relevance sense of support satisfies CE but not SC, while the high probability sense of support satisfies SC but not CE.

It seems that the only serious objection to DS, as a *restricted* notion of support, is the fact (pointed out by Hempel, 1945) that it satisfies the Converse Consequence principle:

(CC)   *Converse Consequence*: If eIh and $b \vdash h$, then eIb.

It is well-known that Converse Consequence cannot be combined with Special Consequence, or with the following Entailment principle (cf., for example, Hesse, 1970a):

(E)   *Entailment*: If $e \vdash h$, then eIh.

Converse Consequence is not implausible as such, however. It has often been thought that evidence supporting a generalization also supports a theory from which that generalization is deducible.[24] Nevertheless, as a general condition of support it seems to be much too liberal. For example, if e supports h, then, by CC, it would support the conjunction $h \& b$ for *any* b whatsoever. Moreover, as Carnap (1950) has shown, CC is not satisfied by the positive relevance criterion of support. Namely, if e is positively relevant to h, then e need not be positively relevant to the conjunction h & b – even if e were positively relevant to b. It can also be easily verified that CC is not satisfied by the high probability criterion of support. It also follows that the measures of corroboration based on positive relevance (cf. Chapter 8.2) do not satisfy CC.

It is interesting to notice that a restricted form of CC is satisfied by the measure $corr_4$ of Chapter 8.3. If we define, for a generalization g and evidence e in a language L, eIg by

$$corr_4(g, e) \geqslant 1 - \varepsilon \quad (0 < \varepsilon < 0.5),$$

then the following principle holds:

> If g and h are generalizations in L, and e is an evidence statement in L, then eIg and $h \vdash g$ imply eIh.

However, this principle does not generally hold in the case that h is a theory in a richer language L'.[25] In other words, if eIg, then eIh holds for all statements h which are logically stronger than g, but still belong to the same language as g. (Here L may be e.g. the observational language $L(\lambda)$, and L' the richer language $L(\lambda \cup \mu)$.)

These results suggest that while an unqualified form of the Converse Consequence Principle is open to serious criticism, some of its modifications might be plausible defended. Brody (1968) has proposed the following principle:

(CC*) If eIh and b explains h, then eIb.

In CC*, the deducibility relation occurring in CC is replaced by a relation of explanation. Brody's suggestion is highly interesting, but rather inconclusive, since he does not have at hand a notion of explanation suitable for principle CC*. Moreover, he explicitly claims that his principle CC* covers cases of non-deductive explanation as well. This part of his suggestion is not very plausible, since the substitution of some inductive relation for 'explains' in CC* easily turns CC* into some unacceptable transitivity principle for inducibility.

What is required in principle CC* is a notion of explanation which is *stronger* than the mere deducibility of h from b.[26] In other words, it is not deducibility as such, but *explanatory deducibility* which is relevant to inductive support. Thus, to make CC* more reasonable than CC itself, it should be required that even if h & b entails h for any b, h & b does not *explain h*. Similarly, even if *h* entails $h \vee b$ for any *b*, *h* does explain $h \vee b$. The notion of deductive explanation developed by Tuomela (cf. Tuomela, 1972 and 1973) has, indeed, these features. If the explanatory relation, call it '*E*', defined by Tuomela's (1973) DE-model or DEL-model of deductive explanation is substituted to CC*, then at least the usual examples showing that the notion of positive relevance cannot satisfy CC are blocked. However, it remains to be studied which probability measures are such that the relation of positive relevance defined by them satisfies CC*.

These remarks are related to Smokler's (1968) discussion of inductive support (or confirmation). Smokler argues that there are two fundamentally different notions of support: one which satisfies Entailment and Special Consequence, but neither Converse Entailment nor Converse Consequence; and another which satisfies Converse Entailment and Converse Consequence, but neither Entailment nor Special Consequence. He suggests that the former notion is characteristic of both 'enumerative' and 'eliminative' inference, and the latter of 'abductive' inference. This interesting suggestion is somewhat vague for the reason that there exists no explicit definition of the inducibility relation *I* that would make *I* satisfy all the principles characteristic of abductive inference. The High Probability Criterion does indeed satisfy the conditions of enumerative inference. However, while the Positive Relevance Criterion satisfies Converse Entailment, it does not generally satisfy Converse Consequence.

It seems reasonable to suggest that abductive inference should be characterized by principle CC* rather than by CC (cf. Tuomela, 1973). Smokler (1968) says himself that within abductive inference it is correct to say that *e* supports *h* if *h* explains *e*. In other words, abductive inference satisfies the following modification of Converse Entailment:

(CE*)   If *hEe*, then *eIh*.

For the DEL-model of deductive explanation, this principle implies neither CC nor CC*, since the explanatory relation *E* is not transitive. However, it may be suggested that abductive inference is definable by the conditions CE* and CC*. It might also satisfy principles SC* and E*, which are obtained from SC and E, respectively, by replacing the relation ⊢ of entailment by the relation *E* of explanation. This can be defended by the following remark: if the explanatory relation *E* is defined by the

DEL-model, the proofs for the impossibility to combine E and CC, CE and SC, and CC and SC, respectively, do not as such apply to E*, CC*, CE*, and SC*. Moreover, it seems that, at least for many reasonable probability measures, the Positive Relevance Criterion satisfies all the conditions E*, CC*, CE*, and SC*.

One might also view the situation by defining a (restricted) notion I* of inductive support by the following strengthened form of the criterion DS:

(11.1) eI*h iff hEe.

Then one would get the result:

(11.2)   Relation I* satisfies principle CE*, but not principles CE, SC, CC, E, SC*, CC*, and E*.[27]

This result shows that the notion of inductive support I* does not satisfy Smokler's conditions for abductive inference. Nonetheless, it may satisfy these conditions in certain important special cases. This may perhaps be taken to indicate that an unqualified form of these conditions is not satisfactory within situations covered by this notion of support. For example, one may suggest that principle CC* is valid only in those cases in which explanation is transitive. In other words, if eI*h, then not any b which explains h is supported by e, but only such a b which also explains e.

The notion of inductive support characterized by the DEL-model of explanation has the interesting feature that it does not generally satisfy the *Equivalence Condition*:

If eIh and $\vdash h \equiv b$, then eIb.

This follows from the fact that the DEL-explanations are not invariant with respect to the substitution of logically equivalent explanantia. To illustrate this, we show how an argument of Rozeboom (1970), p. 95, can be handled.

Rozeboom argues that the Converse Entailment principle cannot be restricted so that if a theory T entails a statement e, then e supports only the conjunction of those conjunctive components of T that are needed to deduce e from T. Let T be any theory such that $T \vdash e$. Then we know that $\vdash T \equiv (e \,\&\, (e \supset T))$. Hence, T can be represented in a logically equivalent form where e occurs as a conjunctive component. Moreover, e is the only component which is needed to deduce e from 'e & $(e \supset T)$'. Consequently, Rozeboom argues, we do not obtain any answer to the question whether e lends support to anything in T over and above e itself. However, suppose that TEe so that eIT by (CE*). The fact that $\vdash T \equiv (e$ & $(e \supset T))$ does not now lead to any untoward consequences. Since neither eEe nor [e & $(e \supset T)$] Ee holds in the DEL-model of explanation, we cannot conclude that e here supports e and only e. Moreover, even if we cannot here conclude that e supports the conjunction 'e & $(e \supset T)$', we still have the fact that e supports T.

Let us summarize the main points of our discussion so far. The notion of inductive support based on converse entailment relations cannot serve as a general definition of support, but may still be reasonably applicable in special situations. The definition DS is too simplified, however. A reasonable way to modify DS is to introduce a relation of explanation in the place of the relation of entailment – either in the Converse Consequence Condition (and in some other conditions as well) or in the definition DS itself.

4.3. Several theorists of induction, advocating the hypothetico-deductive view of scientific inference, have contended that one should be able to account for situations in which the inductive evidence for a scientific statement (theory, law, or singular prediction) is only 'indirect'. A great deal of this discussion has been stimulated by Whewell's idea of the *consilience of inductions*. In his *Novum Organon Renovatum* (1858), Whewell defines this notion as follows:

The Consilience of Inductions takes place when an Induction, obtained from one class of facts, coincides with an Induction, obtained from another different class. This Consilience is a test of the Theory in which it occurs.

This idea is expressible in other terms in the following way. Let g_1 and g_2 be two different generalizations in $L(\lambda)$ which are supported by evidence statements e_1 and e_2, respectively. Then a 'consilience' of g_1 and g_2 takes place when they are both found to be deducible from one and the same theory T in $L(\lambda \cup \mu)$. Thus, theory T establishes deductive systematization between the observational generalizations g_1 and g_2.

Whewell's contention that the consilience is "a test of the theory in which it occurs" corresponds directly to the definition DS of inductive support. However, it has often been thought that through the consilience of the generalizations g_1 and g_2 these initially independent generalizations come to support *each other* – or that the evidence for g_1 comes indirectly to support g_2, and conversely. The following quotations illustrate this view (which does not seem to be Whewell's own)[28]:

After the explanation [by a theory H] L_1, L_2, and L_3 may therefore be more probable than they were before, because each of them derives support indirectly from the evidence in favour of each of the others. This is the consilience of inductions which fit together into a theory. (Kneale, 1949, p. 108.)

When a non-empirical hypothesis thus explains a number of probable generalizations, an important consequence is that each of the latter indirectly evidences and probabilifies every other, so that each is probabilified in a higher degree than it is simply by the direct evidence for it. This relates to what Whewell calls 'the consilience of inductions'. (Day, 1961, p. 60.)

It may happen that [a law] L is jointly derivable with other laws L_1, L_2, etc., from some more general law (or laws) M, so that the direct evidence for these other laws counts as (indirect) evidence for L. (Nagel, 1961, pp. 64–65.)

It is generally considered a consequence of what Whewell called 'consilience' that if two hypotheses, h_1 and h_2, may both be treated as cases or exemplifications of a more general theory, T, then any evidence that raises the degree of confirmation for h_1 and therewith of that of T, must also raise the degree of confirmation for h_2. Moreover, in these circumstances, if h_1 raises the degree of confirmation for T, it also raises that for h_2. (Cohen, 1968a, p. 70.)

The authors of these passages find this idea of consilience acceptable, and think that it can be accounted for by a suitable notion of inductive support. In contrast with them, some authors have claimed that notions of support that do account for such a consilience should be rejected. Thus, Rozeboom (1970), pp. 93–95, claims that the hypothetico-deductive account of theory development "fails abjectly to provide any acceptable standards for the confirmation of theory by data," since it does not sasifsy the following principle:

If C_1 and C_2 are both unverified consequences of theory T, subsequent verification of C_1 necessarily increases the credibility of T as a whole, but need not increase – and in fact may even decrease – the credibility of C_2.

A simple argument suffices to show that all the authors quoted above have misconceived the situation. There is in fact *no* reasonable sense of inductive support that would allow us to infer that g_1 supports g_2 merely by virtue of the consilience, i.e., the fact that both are deductive consequences of some theory T (cf. Hesse, 1970b). The reason for this is the impossibility to combine the principles of Converse Entailment and Special

Consequence – that is, indeed, the same fact from which the transitivity dilemma for Hempel's account for inductive systematization arose. Thus, to infer $g_1 IT$ from $T \vdash g_1$, we need Converse Entailment, and to infer $g_1 Ig_2$ from $g_1 IT$ and $T \vdash g_2$, we need Special Consequence. Thus, we can conclude that if g_1 in fact supports g_2, it cannot do so by virtue of the consilience.

A related argument has been put forward by Putnam (1963) and Hesse (1968) to show that, in any probabilistic confirmation theory, theories (in the richer or even in the same language) cannot increase the confirmation of their deductive predictions. Thus, suppose that g is a law which is deducible from a theory T. Suppose further that e is an evidence statement which does not directly support g, but which supports T. Then e cannot support g indirectly *via* the mediating theory T for the simple reason that the probability of g on e, $P(g/e)$, is a function of g and e only. If e and g are independent, they are independent – irrespective of whether they are deducible from the theory T or not.

If the conclusion of this argument were inescapable, we should be forced to abandon the hope of accounting for the theoretical support of laws within a probabilistic logic of induction – or considerably to revise our notions of a scientific theory and a scientific law. Hesse (1968), has, indeed, chosen the latter route by adopting the view of theories as sets of analogical laws. And, persuaded by Hesse's arguments, Pietarinen has concluded that "it does not seem possible to incorporate the idea of 'indirect' confirmation of laws into the inductive logical framework" (cf. Pietarinen, 1972, pp. 116–118).

Nonetheless, rejection of the possibility of a probabilistic inductive logic for deductive theories seems hasty to us. We have shown above how the theoretical support of laws can be accounted for in those cases in which the relation between the theory and law in question is only inductive. We have thus shown how theories can increase the support of their 'inductive predictions'. Should not the case in which this relation is deductive be even much easier to handle? That this is indeed so, is readily seen if we follow our approach above by taking the theoretical support to be *direct* in the sense that theories are taken to be a part of the accepted evidence for laws, or a part of the 'knowledge situation' in the light of which the laws are evaluated.

To see how this suggestion works, let g be a non-tautological law which is not supported by the evidence e alone. If now $(T \& e) \vdash g$, then $P(g/e \&$ $\& T) = 1$. Therefore, we can conclude that T supports g (relative to e), in the sense of positive relevance or high probability. In fact, this will be the case for any inductive relation that satisfies the Entailment Principle (cf. above). The crucial point here is, of course, that this theoretical support for g is not indirect support by e via T, but is direct support by theory T itself.

That this solution may, at first, seem too simple, reflects only the fact that the support involved in deductive situations is easier to handle than that involved in inductive situations. Yet, this account does not trivialize the problems connected with deductive situations. As we have seen in Chapter 8, our measure $corr_4$ for the degree of corroboration of generalizations g in $L(\lambda)$ on evidence e in $L(\lambda)$ and theories T in $L(\lambda \cup \mu)$ need not be maximal even if g is deducible from e and T. For such theories, measure $corr_4$ takes into account also the eliminative force of T.

It is surprising that this simple solution has not, so far, received more serious consideration. It is trivially true that if two laws g_1 and g_2 are inductively independent, then there is no hope whatsoever of showing, by any kind of 'consilience', that this same g_1 inductively supports the same g_2. The natural way of making sense of such a consilience by a theory seems to be to show that it is g_1 and T together, or g_1 in the light of T, which supports g_2. This is, in short, our strategy in treatment of the inductive role of theories within scientific inference.

One reason which certainly has prevented some authors from accepting theories as evidence is the empiricist doctrine that inductive support should always be based on observational evidence. We have already argued against this view in Section 2. As a technical objection, we may still mention Hesse's (1968), pp. 235–236, claim that the suggestion to accept theories as evidence, and thus to give them probability 1,

does not adequately explicate situations where we wish to compare the c-values of incompatible theories which are confirmed by the same data, for two incompatible theories cannot both have c-value 1.

This argument is beside the point here. If it serves any purpose, we can compare the probabilities of two incompatible theories T_1 and T_2. Whatever the result of this comparison is before the acceptance of one of them as evidence, the probability of T_2 becomes zero if T_1 is accepted as evidence. However, of course, we do not accept both T_1 and T_2 at the same time.

4.4. Let us briefly comment on some proposed solutions to the consilience situation where g_1 and g_2 are deducible from T, and g_1 and g_2 are inductively independent. In view of our discussion, any solution attempting to establish that g_1 alone supports g_2 is unacceptable. For this reason, we reject Cohen's (1968a) solution – which, moreover, is based on a condition of 'Instantial Comparability' that we find unacceptable. This solution is also unnecessarily restrictive, as it applies only to a case where g_1 and g_2 are substitution instances of T. Hesse's (1968) and Mackie's (1968) accounts are based on the notion of analogical inference, which we do not discuss in this work.

Kneale's discussion of consilience (cf. Kneale, 1949, pp. 107–109) is interesting, but it seems that, when his informal examples are formalized, his own interpretation of these examples has to be considerably modified. Kneale's first example is a case in which a biological law is explained deductively by physical and chemical laws:

> When such an explanation has been given, the probability of the biological generalization may very well be greater than it was before. For the biological generalization cannot now be less probable than the physical and chemical laws from which it is seen to follow, and, since these laws, being of greater generality, have presumably been confirmed in many more instances than those which provide evidence for the biological generalization, it is reasonable to suppose that their probability may be greater than that which the biological generalization had attained before the explanation.

Here Kneale seems to have in mind the following idea. Let g be the biological generalization, and let T represent the physical and chemical laws. Suppose further that e is the total observational evidence, and that e' is the part of e which is relevant to g. In other words, we have

(11.3) $P(g/e) = P(g/e')$.

Since $(e \& T) \vdash g$, we have

(11.4) $P(g/e) \geqslant P(T/e)$.

However, if it is now assumed that $P(T/e) > P(g/e')$, formulae (11.3) and (11.4) yield a contradictory conclusion, viz., $P(g/e') > P(g/e')$. Consequently, in contrast with Kneale's idea, we cannot assume that evidence supporting T comes to support g by virtue of the deducibility of g from T.

Another way of formalizing Kneale's example may be suggested. Let

g and e be as above, and let T now be a conjunction T_1 & T_2. Suppose that

(11.5) $(e$ & T_1 & $T_2) \vdash g$

and

(11.6) $P(T_2/e$ & $T_1) \geqslant P(T_2/e) > P(g/e)$.

Since (11.5) entails that

$$P(T_2/e \ \& \ T_1) \leqslant P(g/e \ \& \ T_1),$$

we obtain, by (11.6),

(11.7) $P(g/e) < P(g/e \ \& \ T_1)$.

Formula (11.7) indicates that T_1 is positively relevant to g relative to e. Accordingly, it can be said that T_1 here explains g inductively (cf. Chapter 7), or that the acceptance of T_1 as evidence here gives theoretical support to g (cf. Chapter 8). In other words, T_1 has a *direct* inductive effect upon g.

T_2 plays an interesting role in this example. If T_2 is accepted as part of evidence, then T_1 and T_2 taken together provide a deductive explanation of g. However, T_2 may here also play the role of additional or auxiliary assumptions which are not accepted as part of evidence (as T_1 is). In this case, g is explained inductively by e and T_1, or supported directly by T_1. However, since (11.7) was established on the basis of assumptions (11.5) and (11.6), which involve T_2, T_2 may be said to play an *indirect* role in the evaluation of the inductive support of g.

Kneale's second example is a case in which three laws g_1, g_2, and g_3 are deductively explained by a theory T which is not equivalent to the conjunction g_1 & g_2 & g_3. Here, Kneale argues, T cannot be more probable than g_1, g_2, and g_3 "after they have been explained", but it may be more probable than g_1, g_2, and g_3 "before they were explained." This example can (and should) be reduced to the treatment of the preceding example. However, then the increased support for a law, say g_1, does not come from the other laws, as Kneale thinks, but from the theory in question. Again, our knowledge of this increased support for g_1 may be mediated by theoretical background assumptions.

From the viewpoint of the present work, these examples are of special interest, since they can easily be represented within our framework. In this sense, they also show what can reasonably be included to the proper content of the idea of consilience. More generally, as it was seen above, situations of the following kind can easily be found within our framework. Let evidence e_1 be positively relevant to a generalization g in $L(\lambda)$, and suppose that evidence e_2 is irrelevant to g relative to e_1, i.e.,

$$P(g/e_1 \,\&\, e_2) = P(g/e_1) > P(g).$$

We may then be able to find a theory T in $L(\lambda \cup \mu)$ such that

$$P(g/e_1 \,\&\, e_2 \,\&\, T) > P(g/e_1).$$

Here, e_2 and T together directly support g in the sense of positive relevance. The simple 'consilience situation' in which $(e_1 \,\&\, e_2 \,\&\, T) \vdash g$ is a case in point; another is provided by our treatment of Kneale's example. We may also be able to find a case in which

$$P(g/e_1 \,\&\, e_2 \,\&\, T) > P(g/e_1 \,\&\, T)$$

(supposing that g is not deducible from '$e_1 \,\&\, T$'). In this case, e_2 is seen to support g relative to e_1 and T in the sense of positive relevance. Formally, it is also possible that e_2 is here a law not deducible from T.

These examples can be taken to indicate that theories may play an important inductive role in situations which are not very far from the situations that have been discussed in connection with the 'consilience of inductions'.

5. CONJECTURES

5.1. It is a truism that any attempt aiming at specific results about philosophical problems of induction is relative to some framework and some set of philosophical and technical assumptions. The main part of our technical results is geared to Hintikka's system of inductive logic, but we also have a great number of technical definitions and results which are independent of e.g. specific assumptions on inductive probabilities. We have tried to make as few philosophical assumptions as possible. Therefore, the support we found for scientific realism should be convincing even for an intrumentalistically oriented philosopher.

Let us now recall our technically restrictive assumptions. First we have the assumptions concerning the conceptual framework. When applying Hintikka's system of inductive logic, we operate with monadic first-order languages. This means that we have not dealt with languages containing arbitrary n-place predicates. Neither have we dealt with higher-order languages and intensional languages. Therefore, we have not discussed quantitative theories, such as theories within mathematical physics and probabilistic theories employing random variables. Secondly, we have the assumptions made within Hintikka's inductive logic. As we have frequently pointed out, some of these assumptions are quite inessential. Among these we have the assumption that $\lambda(w) = \lambda$ and the restriction of our treatment to universal statements. However, it is technically possible to extend our approach to the singular case, but then the assumption $\lambda(w) = \lambda$ should be dispensed with. We have normally discussed the cases where only one new predicate is introduced. But, as we showed in Chapter 3, our results can be extended to the cases involving several new predicates.

Some of our other assumptions may be regarded as more restrictive. First, we have dealt only with theories which are of universal form and do not make categorical existential claims. However, our approach can be extended so as to cover theories making existential claims. In fact, in the cases where evidence e realizes the existential claims of a theory T, all probabilities of the form $P(g/e \ \& \ T)$ are given directly by our formulas. In other cases, some modifications are needed. It should be noticed that the cases where the theory is a constituent belong to deductive and not inductive systematization. Another, more serious restriction, is our assumption that the primitive predicates of the language used are inductively symmetric. We shall comment on this problem below. Thirdly, we have assumed that the parameters α and λ do not change when new predicates are introduced to the language. This problem was discussed in Chapter 10.

The assumptions concerning the conceptual framework and inductive logic restrict the field of the methodological situations that our approach covers. Still they do not diminish the philosophical importance of the results which can be proved within our framework. The most central of our results concerning the role of theoretical concepts within inductive systematization have the following character. We have showed that within certain classes of theories (characterized by indices, such as r and b') and relative to certain inductive situations (characterized by parameters α and λ) both evidential and non-evidential theoretical concepts are logically indispensable and desirable for inductive systematization. It is somewhat surprising that such results can be obtained in such a restrictive framework as ours. It seems that when these restrictive assumptions are removed, similar and even much stronger results will be forthcoming.

This conjecture can be supported, for instance, by the following arguments. First, theories employing several theoretical predicates and theories making existential claims are generally epistemologically and logically stronger than the simple theories that we have dealt with. Therefore, it is reasonable to expect them to be inductively stronger as well. Secondly, if parameter values could be freely changed, then naturally inductive systematization would be easier to establish. Thirdly, as has been shown in detail by Hintikka and Tuomela (1970), Hintikka (1973), and Tuomela (1973), the notion of *depth*[29] of theories becomes central to the gains due to theoretical concepts within polyadic (relational) first-order languages. These authors have shown it to be the case within deductive systematization. Moreover, Hintikka and Tuomela (1970), p. 315, have suggested that the gains due to theoretical concepts are connected with the inductive support which theories receive from the deductive systematization of observational statements: the decrease of the depth of a theory due to the introduction of theoretical concepts increases the degree of testability of the theory. Analogously with this, we conjecture that the notion of depth plays a central role within hypothetico-inductive inference performed in polyadic languages.

5.2. One common feature in the best known systems of inductive logic, such as Carnap's and Hintikka's, is that at least in their original forms the primitive extralogical concepts (predicates) are taken to be inductively symmetric. However, in so far as a system of inductive logic can be considered to reconstruct a fragment of scientific or of natural language used within induction, situations in which all the primitive predicates cannot be regarded as being on *a par* with respect to confirmation may arise. Goodman's paradox provides a case in point. The existence of this kind of problematic cases has to be duly acknowledged, and no full-blown system of inductive logic can be regarded as satisfactory, unless it can at least technically account for such inductive asymmetries.

Within Hintikka-type inductive logics basically four different approaches to account for the asymmetry of predicates have been taken.

(1) Differences in the values of the regularity parameter α may be construed to explicate inductive asymmetries between systems of predicates

in the manner of our first (partial) solution to Goodman's paradox (cf. Chapter 10.3). (An analogous remark applies also to the parameter λ.)

(2) Carnap has tried to account for inductive asymmetries between predicates in terms of their inductive *irrelevance* and relevance (cf. Carnap and Stegmüller, 1959). He introduced a new parameter to measure such irrelevance. As has been shown by Hesse (1964), it is possible to give an account of singular *analogical* inference by means of this parameter. Pietarinen (1972) has applied Carnap's idea to Hintikka's system of inductive logic by introducing a parameter η to measure the irrelevance of families of predicates. This enables him to explicate the intuitive inductive difference between the problematic generalizations g_G and g_R in Goodman's paradox.

(3) Hintikka (1969a) has technically created asymmetries by so *ordering* the primitive predicates that the Ct-predicates are thought of as arising through a number of successive dichotomies. As a result, he is able to create confirmation functions that are *selective* in the sense of making some predicates more sensitive to confirmation by instances than some others. Hintikka has applied his results to discuss both Hempel's and Goodman's paradox.

(4) A background theory T which introduces new concepts may have the effect that the primitive predicates become inductively asymmetric, as required. This alternative was discussed in Chapter 10.3.

What is common to the above approaches (1)–(4) is that they are based on assumptions concerning the properties of predicates – single predicates or, as in cases (1)–(3), systems of predicates. These assumptions express various *presuppositions of induction* that cannot be stated in the original language, but require stronger languages. Generally it is most natural to regard them as higher-order, at least second-order, languages, even if a first-order language with theoretical predicates may sometimes be sufficient.

Assumptions from which inductive asymmetries are derived should be philosophically illuminating and well-founded. We claimed in Chapter 10.3 that even if Hintikka (1969a) and Pietarinen (1972) technically avoid the paradoxes in their treatments, their solutions still have to be complemented by philosophical explanations. It seems to us that approach (4), at least when it can be carried out within our framework, is preferable to the other alternatives, since it gives a more detailed analysis of the source of asymmetries. The basic idea of this approach is to try to make explicit the various conceptual and factual assumptions concerning the primitive predicates of the system. The resulting background theory, if one succeeds in finding one, will often be a second- or higher-order theory.

Consequently, in a fuller account, at least a second-order inductive logic seems to be needed. We conjecture that:

(a) By means of higher-order theories one will be able to *qualitatively* accomplish the same as by means of the approaches (1)–(3).

(b) In many cases, such higher-order theories have first-order consequences. Such cases can be handled even within our present system of inductive logic. For instance, Goodman's paradox can be technically approached by means of first-order background assumptions, as we did in Chapter 10.3.

(c) By means of background theories we may be able to inductively differentiate between different uses of predicates, such that the same predicates may be considered inductively irrelevant (or more or less so) when they occur in a certain hypothesis, whereas they are highly relevant when they occur in another hypothesis belonging to the same conceptual system. This is not possible within the approaches (1) and (2), since the values of the parameters α and η effect the whole system of predicates.

Note finally that approach (4) is not necessarily exclusive to the other approaches, but may be used to complement them. This is what we at least conjecture, even though a fair amount of technical work is needed to carry out this suggestion.

As we have remarked above, there are a number of interesting philosophical problems which can be related to the question of the inductive asymmetry of predicates. Among them are the general problem of inference by *analogy* and the problem of *lawlikeness* of generalizations. We cannot here go into these problems in any detail, even if we conjecture that the use of new theoretical concepts and background theories will again help in the search for a philosophically interesting solution.

5.3. In classifying statements as lawlike, there is a number of aspects that are relevant: Syntactical, semantical, systematical, methodological, and inductive.[30] A full account of lawlikeness should do justice to all of these aspects. It seems to us that such an account can be given within the framework of scientific realism using the ideas and results of this work. Roughly speaking, such an account might run as follows. By *regularities* we mean general (uniform) dispositional properties that the world may have. These properties are manifested, but not exhausted, by 'constant conjunctions' of events (or facts, etc.).[31] Generalizations which express such regularities are *lawlike*. As our knowledge of the world essentially depends upon conceptual frameworks and scientific theories, whether an (observational or non-observational) generalization is lawlike comes to depend upon its systematic, deductive and inductive, connections to a broader theory or theories. Such theories normally contain deeper explanatory concepts. One may then say that a lawlike generalization can be regarded as a *law* to the extent we can regard the whole broader theory as true (or approximatively true).

Here we have to restrict ourselves to some remarks on the connection between the systemical and inductive aspects of lawlikeness. To bring out one aspect of the situation, we may start from the following famous statement of John Stuart Mill (cf. Mill, 1843, Book III, Ch. 3):

> Why is a single instance, in some cases, sufficient for a complete induction, while, in others, myriads of concurring instances, without a single exception known or presumed, go such a very little way towards establishing a universal proposition? Whoever can answer this question knows more of the philosophy of logic than the wisest of the ancients and has solved the problem of induction.

What Mill has in mind in the above quotation is the difference between the status of a generalization about the chemical properties of a newly-discovered substance, based on the examination of a single piece of this substance, and that of another one stating that all crows are black. This suggests that an important indicator of lawlikeness is the rate at which positive instances increase the probabilities of generalizations.

Within Hintikka's system of inductive logic this rate depends on the parameter α. In view of this, Hintikka (1969b) and Pietarinen (1972) have suggested that α can be used to measure the *degree of lawlikeness* of generalizations. However, this has the consequence that all generalizations within the same conceptual system become equal as to their degrees of lawlikeness, since the parameter α is a characteristic of the *whole* conceptual system. This consequence is not generally acceptable, since, within one and the same conceptual system, there are often generalizations that intuitively differ as to their degree of lawlikeness. Therefore, this approach has to be complemented so that such cases can be adequately dealt with.

Our second solution to Goodman's paradox suggests that this can be accomplished by means of background theories introducing new theoretical concepts. (Thus, in Chapter 10.3, generalization g_G may be said to have a greater degree of lawlikeness than generalization g_R.) This solution can be seen to work nicely in the case that Mill had in mind. As an illustration, an example of Kyburg (1970), pp. 121–123, can be considered.

Let $F(W)$ mean that W is a species of fish, and let $B(x)$ mean that x is a pair which under normal conditions exhibits a certain mating pattern. Let further T' be the second-order theory

$$T' = (W)[F(W) \supset ((\exists x)(W(x) \& B(x)) \supset (x)(W(x) \supset B(x)))].$$

T' may be a consequence of a broader biological theory containing characteristic theoretical concepts. If S is a certain particular species of fish, say salmon, then T' has as its first-order consequence the following theory T:

$$T = (\exists x)(S(x) \& B(x)) \supset (x)(S(x) \supset B(x)).$$

This theory states that if one normal pair of salmons, under normal conditions, exhibits a certain mating pattern, then all normal pairs of salmons will, under normal conditions, exhibit that same behavior.

Let e_n be evidence containing n substitution instances of generalization $g = (x)(S(x) \supset B(x))$. Since $(T \& e_1) \vdash g$, we have $P(g/e_1 \& T) = 1$. Thus, in the light of theory T, already *one* positive instance is sufficient to raise the probability of g to its maximum value. (Note that $P(g/T) < 1$.) However, $P(g/e_n)$ may be much smaller than one even if n is relatively large.

The role of theoretical concepts in this example is somewhat indirect. They are used in the broader biological theory, and, as Kyburg remarks, may include theoretical concepts designating innate properties. This broader theory has T' and T as its deductive consequences. Here T guarantees that g has a maximal rate of growth *vis-à-vis* probabilities, and thus indicates that g is a highly lawlike generalization. In other words, g receives theoretical support from the biological theory, which itself may be acceptable in virtue of its systematic, deductive or inductive, connections with other theories, and of its explanatory relation to other observational generalizations. In some lucky cases, it may also get direct support from evidential commerce with the world.

This example may perhaps be considered too special, as the theory T

is unusually strong. However, similar examples are forthcoming in all the cases where a theory gives *theoretical support* to an observational generalization. In Chapter 8, we have systematically discussed the cases where a theory gives theoretical support to a generalization either in the positive relevance sense ($corr_1$, $corr_2$, and $corr_3$) or in the sense of $corr_4$. A striking example is provided by $corr_4$ for which we may have, for a generalization g, evidence e, and a theory T, $corr_4(g/e) \to 0$ and $corr_4(g/e \ \& \ T) \to 1$. This, and similar examples, show that theories can strongly effect the rate of growth of the support generalizations may receive, and hence increase their degrees of lawlikeness.

As a special case of the role of theoretical support, one can mention *contrafactual conditionals*. For example, such contrafactual conditional generalizations which have no instances (e.g., statements employing highly idealized concepts), or no observed instances, may still have high a degree of theoretical support due to the systemicity factor. This may be taken to indicate that these generalizations can nevertheless be lawlike.

These remarks suffice to indicate the relevance and importance of the systemicity factor in analysing the concept of lawlikeness. Against the opinion of such critics of the systemicity view of lawlikeness as Pietarinen (1972), they show that, also in this connection, the role of theoretical concepts and theories should and can be accounted for within the framework of inductive logic.

NOTES

[1] This emphasis is so common that it may suffice here to refer to only one of its advocates. Campbell (1921) argues that we are to exclude every particular event from the subject-matter of science, since science is a study of those judgements concerning which universal agreement can be obtained.

[2] An interesting discussion of the developments of the idea that science is a search of 'invariant' features of the world can be found in Kaila (1939).

[3] Hintikka (1969c) suggests that there is a close analogy between 'analysis' in the sense of Greek geometers and 'analysis' in the sense which the founders of modern natural science (such as Galilei, Descartes, and Newton) used to describe their own experimental method. Since geometric analysis can be shown to be a non-calculable or non-algorithmic process, this analogy suggests that this also holds for experimental analysis. Hintikka also points out that in the cases where (in addition to mere analysis) new concepts are introduced the old (perhaps observational) concepts are typically definable in terms of the new (theoretical) concepts.

[4] See also the interesting historical remarks in von Wright (1957), especially pp. 206–208.

[5] See our discussion on Popper in Chapter 8 and in Section 1.

[6] Note, however, that the derivation of probabilistic statements from other such statements is often deductive (assuming the calculus of probability), and belongs to deductive-probabilistic rather than inductive systematization (cf. e.g. Niiniluoto, 1973a).

[7] This refutes the claims of Bohnert (1968), Hooker (1968), and Stegmüller (1970) that all theories which do not establish deductive systematization among observational statements are 'empirically trivial' and cannot achieve inductive systematization. This claim is discussed in Chapter 9 above, and in Niiniluoto (1973b).

[8] Hempel's specific example is

$$(x)[(y)R_1(x, y) \supset (Ez)R_2(x, z)].$$

where R_1 and R_2 are observational two-place predicates. Here, the prediction is to be based upon the truth of the antecedent $(y)R_1(a, y)$, for some a, which can be established by a finite observation report only *inductively*.

[9] Canfield and Lehrer (1961) have tried to argue that the inference from a law and the initial conditions to a prediction cannot be deductive. That their argument is incorrect has been shown by Coffa (1968). See also Stegmüller (1969), pp. 143–153.

[10] That induction cannot be transitive is contended by many philosophers. Striking counter-examples of the transitivity principle can be found in Salmon (1965). Even modified forms of the transitivity principle can be questioned (cf. the discussion in Niiniluoto, 1973a). In view of this, it is difficult to see how one could justify, for example, Nicod's (1930) ideas about 'secondary induction' or Day's (1961) account of 'derivative induction'; they assume that induction has a kind of cumulative effect, so that the conclusions of one induction can be used as premises of a new induction. The inductive rule IR of Lehrer (1970) is similarly based on this idea, and has been critized in Niiniluoto (1971 and 1973a).

[11] The fundamental point of Hintikka's criticism is as follows. Suppose that we evaluate the reliability of strong generalizations by means of the degrees of instance confirmation of their universal parts. If now the universe and the evidence are sufficiently large, the more Ct-predicates a strong generalization contains, the more reliable it is. In particular, the least reliable generalization is that corresponding to the constituent C_c, which nevertheless seems to be the one which the evidence favors most. A similar argument can also be used against Hesse's (1969b) suggestion of restricting the confirmation of laws (within Carnap's system) to generalizations in *finite* domains of individuals.

[12] The assumption that generalizations on infinite domains receive the probability zero has been defended in several ways. For example, if probabilities are interpreted as betting quotients, one can ask whether it is reasonable to assign a positive betting quotient to a generalization. It has thus been claimed that it does not make sense to bet on generalizations, or to bet on them with non-zero odds (cf. Hacking, 1965, p. 23; Stegmüller, 1971, pp. 15, 59). These arguments have been questioned in Hintikka (1971b) and Pietarinen (1972). Popper wrongly believes that he has proved $P(g) = 0$, where g is a universal law, for *all* acceptable probability measures. (Cf. Popper, 1959, appendices vii and viii.) His argument runs essentially as follows. Suppose that $g = (x)M(x)$. Then

$$P(g) = \lim_{n \to \infty} P(M(a_1) \, \& \, ... \, \& \, M(a_n)).$$

If now (1) $P(M(a_i)) = P(M(a_j)) = p < 1$, and (2) $P(M(a_i) \, \& \, M(a_j)) = P(M(a_i))P(M(a_j))$, for all $i \neq j$, then

$$P(g) = \lim_{n \to \infty} p^n = 0.$$

The crucial step in this argument is the assumption (2) of the probabilistic independence of all the singular statements $M(a_i)$ and $M(a_j)$, when $i \neq j$. For measures P for which

(∗) $P(M(a_n)/M(a_1) \ \& \ ... \ \& \ M(a_{n-1})) > P(M(a_n))$

may hold, Popper's proof does not remain valid. Popper excludes (∗) as an unacceptable "synthetic *a priori* principle of induction," and therefore rejects all measures P satisfying (∗). But he does not tell us why his principle has a different status from that of (∗). Indeed, if Popper is right, it seems that the independence assumption (2) too should be classed among the 'synthetic a priori principles'. Consequently either Popper's argument is too weak to exclude any measure P, or otherwise it excludes all of them.

13 Thus, Popper (1959), p. 376, adopts the *ad hoc* principle that if, for generalizations h_1 and h_2, *cont* $(h_1) > cont$ (h_2) for all sufficiently large universes then one can assume that, in an infinite universe, the content of h_1 is greater than that of h_2 in the sense of the *fine structure* of content – even if there *cont* $(h_1) = cont$ $(h_2) = 1$. See also the critical remarks in Pietarinen (1972), p. 76.

14 See Hintikka and Hilpinen (1966), pp. 15–19; Hilpinen (1968), pp. 68–76; Hilpinen and Hintikka (1971), pp. 304–306.

15 Hesse (1969b) has challenged Nagel's view, quoted above, by suggesting that it is sufficient to have confirmation values for generalizations within limited domains. This is not very plausible if observational evidence statements can be theory-laden. In note 11, we pointed out another difficulty which Hesse has to face. Moreover, her defence of the assumption of zero probabilities of generalizations in infinite universes is based on an 'atomistic' intuition which we found unacceptable above. According to such an intuition, it is not reasonable to believe that genuine universal laws have any change of being true, but, on the contrary, we have the best of possible reasons for believing that a counterinstance for a law exists. This view has been critized also in Pietarinen (1972), pp. 74–76.

16 See, for example, Carnap (1963), pp. 987–989, 992; (1971b), pp. 49–52; Lakatos (1968), pp. 407–408; and Rozeboom (1970), p. 96.

17 To take a specific example, suppose that $\lambda = \{O\}$ and $\mu = \{P\}$, where P is a non-evidential theoretical predicate. Let g be the theoretical hypothesis $(x)P(x)$, and suppose that we have observed n individuals all of them having the property 'O'. Then $P(g/e)$ approaches the value $1/(\alpha + 3)$ when $n \to \infty$. This value is always less than or equal to $\frac{1}{3}$, that is, it is never high, which seems reasonable. When $\alpha = 0$, we have $P(g) = \frac{1}{4}$ and $P(g/e) = \frac{1}{3}$, so that e is positively relevant to g. If this is found nonplausible, it may taken to reflect the 'overoptimistic' nature of small values of α. Another way of expressing this is the following. If $\alpha = 0$, our universe is estimated to be maximally regular. If all individuals are found to satisfy one and the same predicate 'O' the probability of the generalization $(x)P(x)$ (and, by symmetry, of $(x) \sim P(x)$ as well) is mildly increased. However, if we have observed n_1 individuals satisfying 'O', and n_2 individuals satisfying '$\sim O$', then $P(g/e)$ approaches the value $1/(5 + 2\alpha)$ when $n_1 \to \infty$ and $0 > n_2 > \infty$. In this case, evidence e does not increase the probability of g even if $\alpha = 0$. – If this example is found to be unrealistic, one may easily construct examples where some evidence e in $L(\lambda)$ is positively relevant to a hypothesis containing *both* observational and (non-evidential) theoretical predicates.

18 Barker's related claim that 'transcendent hypothesis of the first kind', i.e., hypotheses

(in $L(\lambda)$) about the existence of unobserved individuals, has been refuted by Hilpinen (1971).

[19] Attempts to base criteria of 'empirical meaningfulness' or 'cognitive significance' on the above methodological notion of testability are unsatisfactory, as they are based on questionable assumptions of the existence of a unique class of 'purely observational' statements that are, as such, 'meaningful'. Such an assumption involves the so called thesis of semantical empiricism, and is unacceptable to a scientific realist (cf., for example, Tuomela, 1973). Moreover, Ayer's (1946) definition of meaningful statements has been shown to be unsatisfactory for technical reasons as well (cf. Church, 1949; Hempel, 1965b). It should be noted, however, that Church's ingenious counterargument to Ayer's definition does not apply to the notion testability based on DS. Popper (1959) has defined 'scientific' or 'non-metaphysical' statements as those which are testable (essentially) in the above sense. This definition has the following consequence: if T is testable, then T & H is testable for *any* H. Popper accepts this: he claims that "empirical theories (such as Newton's) may contain 'metaphysical' elements." However, these elements can be eliminated if we succeed in presenting the theory as a conjunction of a testable and a non-testable part (cf. Popper, 1959, p. 85). But even if this fact were tolerated, Popper's criterion would have the unsatisfactory consequence that all existential statements and, in particular, negations of universal theories would be 'metaphysical' (cf. Hempel, 1965b). Thus, Popper's criterion seems to be too restrictive. Other reasons for this conclusion are given below.

[20] Let $T = (x)(Fx \supset Gx)$ be a theory in $L(\lambda)$, and let $e = (Fa \supset Ga) \supset Ha$ and $h = Ha$ be statements of $L(\lambda)$ such that condition (c) of DS is satisfied. Then T, e, and h satisfy the definition DS. Stegmüller finds this unacceptable, because h "does not have anything to do with" T. However, DS implies here that it is not h alone but $\{e, h\}$, i.e., e and h together, which supports T. Furthermore, it is not true that e *and* h together do not have anything to do with T.

[21] In contrast with Stegmüller, we have required (in condition (b) of DS) that T is essential for the deduction of h, i.e., that h is not deducible from e alone.

[22] Stegmüller claims that statements $T = (x)(y)[\sim (Rxy \& Ryx) \supset (Rxy \& Ryx)]$, $e = Rab$, and $h = \sim Rba$ satisfy the definition DS. If so, $\{e, h\}$ supports T. But T is logically equivalent to statement $(x)(y)Rxy$ which cannot be supported by $\{e, h\}$, since the test statement h contradicts it. This argument is mistaken, however, for the reason that $\{e, h\}$ does not support T. In the first place, h is not deducible from e & T. In the second place, even though e is here deducible from h & T, $\{h, e\}$ cannot support T for the simple reason that h is not compatible with T and, consequently, condition (a) of DS is not satisfied.

[23] In this spirit, Rozeboom (1970), p. 97, has remarked that "two scientific propositions P and C are often so related that verification of C increases the credibility of P even if C is *not* a logical consequence of P or conversely."

[24] See, for example, Nagel (1961), pp. 64–65, and Brody (1968).

[25] This follows from the fact that if g and h are generalizations in $L(\lambda)$ such that $h \vdash g$, then all constituents belonging to the distributive normal form of h belong also to the normal form of g. Accordingly, $corr_4(h, e)$, that is, the minimum of $P(C_i/e)$ for constituents C_i in the normal form of h, is greater than $corr_4(g, e)$. This need not hold, however, when h and g belong to different languages.

[26] Somewhat curiously, Rozeboom (1970), p. 97, says that *inductivists* are "prepared to find that the relation between an inductively justified conclusion and its evidence depends upon logical structure a great deal more than mere entailment."

[27] That I^* satisfies CE^* is trivially true by (11.1). That it does not generally satisfy SC, SC^*, CC, and CC^* follows from the fact that E is not transitive. (It is not intransitive, however.) It does not satisfy E^* since E is not symmetric, and it does not satisfy CE since $h \vdash e$ does not imply hEe.

[28] For other discussions of the consilience of inductions, see Hesse (1968 and 1970b), Cohen (1968b and 1970), and Mackie (1968).

[29] The depth of a sentence can be defined to be the maximum number of nested quantifiers occurring in it. Intuitively, it corresponds to the number of individuals considered in their relation to each other in the deepest part of the sentence.

[30] For the syntactical aspects concerning the logical form of lawlike statements, see Nagel (1961), pp. 57–59, and Hempel (1965a), p. 340. For the semantical aspects concerned with what lawlike generalizations assert, see Chisholm (1955), Goodman (1955), Nagel (1961), Rescher (1969), Stalnaker and Thomason (1970), and von Wright (1971). For the systematical aspects concerned with 'systemicity', i.e., the systematic connections of lawlike statements with a larger theoretical framework, see Braithwaite (1953), pp. 293–318, Nagel (1961), pp. 47–78, Bunge (1963), pp. 153–179, and Hempel (1966), pp. 54–58. By the methodological aspects we mean such requirements as the explanatory power of lawlike statements (cf. the works referred to above in connection with the systemicity factor). By inductive aspects we mean questions of the confirmation and acceptance of lawlike generalizations, and of the support they can lend to other statements (such as counterfactual conditionals). For discussions dealing with these questions, see Braithwaite (1953), Goodman (1955), Nagel (1961), Hintikka (1969b), von Wright (1971), Pietarinen (1972), and Uchii (1972).

[31] It should be emphasized that we do not here advocate the so-called constant-conjunction view of lawfulness. We agree with Rescher (1969) that lawfulness manifests ifself through *hypothetical force*: a lawlike generalization does not state only that, say, all X's are Y's, but rather that all X's are Y's *and* if a were an X (though it is not), then it would be an Y. In accordance with this, we have characterized the objective regularities that lawlike generalizations express as general dispositional properties of the world. This means that lawfulness can never be wholly established by observation. However, this is a general feature of all genuinely theoretical knowledge of the world, and need not make lawfulness *mind-dependent*, even if Rescher (1969) argues so. In other words, lawfulness is no more or less dependent upon minds than theoretical knowledge expressed within a conceptual system. It is another question how the way of expressing lawlike and accidental generalizations, respectively, should differ from each other. One way of viewing it is to introduce *modal* operators to the object language in which the generalization is expressed (cf. von Wright, 1971; Uchii, 1972), but this need not be the only way out. For example, Braithwaite (1953), p. 316, and Nagel (1961), p. 72, suggest that the hypothetical or counterfactual force of the conditional 'If p, then q' could be taken into account by interpreting it as an implicit *metalinguistic* statement asserting that there is a scientific theory T such that $T \vdash \sim (p \,\&\, \sim q)$ holds independently of the eventual falsity of p.

BIBLIOGRAPHY

Achinstein, P., *Concepts of Science, A Philosophical Analysis*, The Johns Hopkins Press, Baltimore, 1968.

Ackermann, R., 'Inductive Simplicity', *Philosophy of Science* **28** (1961) 152–160.

Aristotle, *Posterior Analytics* (translated into English in *The Works of Aristotle*, ed. by D. W. Ross, Oxford 1928).

Ayer, A. J., *Language, Truth and Logic*, 2nd edition, Victor Gollancz Ltd, London, 1946.

Bacon, Fr., *The New Organon*, 1620. (Reprinted in *The Works of Francis Bacon*, ed. by J. Spedding, R. L. Ellis, and D. D. Heath, London 1857–58).

Bar-Hillel, Y., *Language and Information*, Addison-Wesley and the Jerusalem Academic Press, Reading, Mass., and Jerusalem, 1964.

Bar-Hillel, Y., 'Comments on "Degree of Confirmation" by Professor K. R. Popper', *British Journal for the Philosophy of Science* **6** (1955) 155–157.

Barker, S. F., *Induction and Hypothesis: A Study in the Logic of Confirmation*, Cornell University Press, Ithaca, 1957.

Barker, S. F., 'Comments on Salmon's "Vindication of Induction"' in *Current Issues in the Philosophy of Science* (ed. by H. Feigl and G. Maxwell), pp. 257–260, Holt, Rinehart and Winston, New York, 1961. (Referred to as (1961a).)

Barker, S. F., 'The Role of Simplicity in Explanation', in *Current Issues in the Philosophy of Science* (ed. by H. Feigl and G. Maxwell), pp. 265–273, Holt, Rinehart and Winston, New York, 1961. (Referred to as (1961b).)

Barker, S. F. and Achinstein, P., 'On the New Riddle of Induction', *Philosophical Review* **69** (1960) 511–522.

Bohnert, H., *The Interpretation of a Theory*, Ph.D. Dissertation, University of Pennsylvania, Pennsylvania, 1961.

Bohnert, H., 'Communication by Ramsey-Sentence Clause', *Philosophy of Science* **34** (1967) 341–347.

Bohnert, H., 'In Defence of Ramsey's Elimination Method', *Journal of Philosophy* **65** (1968) 275–281.

Braithwaite, R. B., *Scientific Explanation*, Cambridge University Press, Cambridge, 1953.

Brody, B. A., 'Confirmation and Explanation', *Journal of Philosophy* **65** (1968) 282–299.

Bunge, M., 'The Weight of Simplicity in the Construction and Assaying of Scientific Theories', *Philosophy of Science* **28** (1961) 120–149.

Bunge, M., *The Myth of Simplicity*, Prentice Hall, Englewood Cliffs, 1963.

Bunge, M., 'What are Physical Theories About', *American Philosophical Quarterly*, (Monograph Series) **3** (1969) 61–91.

Campbell, N., *What is Science?*, Dover, New York, 1921.

Canfield, J. and Lehrer, K., 'A Note on Prediction and Deduction', *Philosophy of Science* **28** (1961) 204–211.

Carnap, R., *The Logical Foundations of Probability*, The University of Chicago Press, 1950.

Carnap, R., *The Continuum of Inductive Methods*, The University of Chicago Press, Chicago, 1952. (Referred to as (1952a)).

Carnap, R., 'Meaning Postulates', *Philosophical Studies* 3 (1952) 65–73. (Referred to as (1952b).)

Carnap, R., *Meaning and Necessity: A Study in Semantics and Modal Logic*, Chicago University Press, Chicago, 1956.

Carnap, R., 'Preface' to the 2nd Edition of Carnap (1950). (Referred to as (1962a).)

Carnap, R., 'The Aim of Inductive Logic', in *Logic, Methodology and Philosophy of Science: Proceedings of the 1960 International Congress* (ed. by E. Nagel, P. Suppes, and A. Tarski), pp. 303–318, Stanford University Press, Stanford 1962. (Referred to as (1962b).)

Carnap, R., 'Replies and Systematic Expositions', in *The Philosophy of Rudolf Carnap* (ed. by P. A. Schilpp), pp. 966–998, Open Court, La Salle, 1963.

Carnap, R., *Philosophical Foundations of Physics* (ed. by M. Gardner), Basic Books, New York, 1966.

Carnap, R., 'Inductive Logic and Rational Decisions', in *Studies in Inductive Logic and Probability*, vol. 1 (ed. by R. Carnap and R. C. Jeffrey), pp. 5–31, University of California Press, Berkeley, 1971. (Referred to as (1971a).)

Carnap, R., 'A Basic System of Inductive Logic; Part 1', in *Studies in Inductive Logic and Probability*, vol. 1 (ed. by R. Carnap and R. C. Jeffrey), pp. 33–165. (Referred to as (1971b).)

Carnap, R. and Bar-Hillel, Y., 'An Outline of a Theory of Semantic Information', *Technical Report No.* 247, MIT Research Laboratory in Electronics 1952. (Reprinted in Bar-Hillel (1964).)

Carnap, R. and Stegmüller, W., *Inductive Logik und Wahrscheinlichkeit*, Springer-Verlag, Wien, 1959.

Chisholm, R., 'Law Statements and Counterfactual Inference', *Analysis* 15 (1955) 97–105.

Church, A., Review of Ayer (1946), *Journal of Symbolic Logic* 14 (1949) 52–53.

Coffa, J., 'Deductive Predictions', *Philosophy of Science* 35 (1968) 279–283.

Cohen, L. J., 'A Note on Consilience', *British Journal for the Philosophy of Science* 19 (1968) 70–71. (Referred to as (1968a).)

Cohen, L. J., 'An Argument that Confirmation Functors for Consilience are Empirical Hypotheses', in *The Problem of Inductive Logic* (ed. by I. Lakatos), pp. 247–250, North-Holland, Amsterdam 1968. (Referred to as (1968b).)

Cohen, L. J., *The Implications of Induction*, Methuen & Co Ltd, London, 1970.

Cornman, J. W., 'Craig's Theorem, Ramsey-Sentences, and Scientific Instrumentalism', *Synthese* 25 (1972) 82–128.

Craig, W., 'On Axiomatizability within a System', *Journal of Symbolic Logic* 18 (1953) 30–32.

Craig, W., 'Replacement of Auxiliary Expressions', *Philosophical Review* 65 (1956) 38–55.

Craig, W., 'Bases for First-Order Theories and Subtheories', *Journal of Symbolic Logic* 25 (1960) 97–142.

Day, J. P., *Inductive Probability*, Routledge and Kegan Paul, London, 1961.

Feyerabend, P., 'Problems of Empiricism', in *Beyond the Edge of Certainty* (ed. by R. G. Golodny), pp. 145–260, Prentice-Hall, Englewood Cliffs, 1965.

Finetti, B. de, 'Foresight: Its Logical Laws, Its Subjective Sources', in *Studies in Subjective Probability* (ed. by H. E. Kyburg and H. Smokler), pp. 93–158, John

Wiley and Sons, New York, 1964. (English translation of B. de Finetti, 'La Prevision: Ses Lois Logiques, Ses Sources Subjectives', *Annales de l'Institut Henri Poincaré* **7** (1937) 1–68.)

Good, I. J., 'Weight of Evidence, Corroboration, Explanatory Power, Information and the Utility of Experiments', *Journal of Royal Statistical Society* B **22** (1960) 319–331.

Goodman, N., *Fact, Fiction, and Forecast*, Harvard University Press, Cambridge, 1955.

Goodman, N., 'Recent Developments in the Theory of Simplicity', *Philosophy and Phenomenological Research* **19** (1959) 429–446.

Greeno, J. G., 'Evaluation of Statistical Hypotheses Using Information Transmitted', *Philosophy of Science* **37** (1970) 279–293.

Greeno, J. G., 'Theoretical Entities in Statistical Explanation', in *Boston Studies in the Philosophy of Science*, vol. VIII, pp. 3–26, D. Reidel, Dordrecht, 1971.

Hacking, I., *Logic of Statistical Inference*, Cambridge University Press, Cambridge, 1965. (Referred to as (1965a).)

Hacking, I., 'Salmon's Vindication of Induction', *Journal of Philosophy* **62** (1965) 260–266. (Referred to as (1965b).)

Hacking, I., 'Linguistically Invariant Inductive Logic', *Synthese* **20** (1969) 25–47.

Hanna, J., 'A New Approach to the Formulation and Testing of Learning Models', *Synthese* **16** (1966) 344–380.

Hanson, N. R., *Patterns of Discovery*, Cambridge University Press, Cambridge 1958.

Hanson, N. R., 'Is there a Logic of Discovery', in *Current Issues in the Philosophy of Science* (ed. by H. Feigl and G. Maxwell), pp. 91–100, Holt, Rinehart, and Winston, New York, 1961.

Hempel, C. G., 'Studies in the Logic of Confirmation', *Mind* **54** (1945) 1–26, 97–121. (Reprinted in C. G. Hempel, *Aspects of Scientific Explanation and Other Essays in the Philosophy of Science*, pp. 3–46, The Free Press, New York, 1965.)

Hempel, C. G., 'The Theoretician's Dilemma', in *Minnesota Studies in the Philosophy of Science*, vol. 2 (ed. by H. Feigl, M. Scriven, and G. Maxwell), pp. 37–98, University of Minneapolis Press, Minneapolis 1958. (Reprinted in *Aspects of Scientific Explanation*, pp. 173–226.)

Hempel, C. G., 'Inductive Inconsistencies', *Synthese* **12** (1960) 439–69. (Reprinted in *Aspect of Scientific Explanation*, pp. 53–79.)

Hempel, C. G., 'Deductive-Nomological vs. Statistical Explanation', in *Minnesota Studies in the Philosophy of Science*, vol. III (ed. by H. Feigl and G. Maxwell), pp. 98–169, University of Minnesota Press, Minnesota, 1962.

Hempel, C. G., 'Implications of Carnap's Work for the Philosophy of Science', in *The Philosophy of Rudolf Carnap* (ed. by P. A. Schilpp), pp. 685–709, Open Court, La Salle, 1963.

Hempel, C. G., 'Aspects of Scientific Explanation', in *Aspects of Scientific Explanation and Other Essays in the Philosophy of Science*, pp. 331–496, The Free Press, New York, 1965. (Referred to as (1965a).)

Hempel, C. G., 'Empiricist Criteria of Cognitive Significance: Problems and Changes', in *Aspects of Scientific Explanation*, pp. 101–119. (Referred to as (1965b).)

Hempel, C. G., *Philosophy of Natural Science*. Prentice-Hall, Englewood Cliffs, 1966.

Hempel, C. G. and Oppenheim, P., 'Studies in the Logic of Explanation', *Philosophy of Science* **15** (1948) 135–175. (Reprinted in *Aspects of Scientific Explanation*, pp. 245–290.)

Heyerdahl, T., *The Ra Expeditions*, George Allen & Unwin, London, 1971.

Hesse, M. B., 'Analogy and Confirmation Theory', *Philosophy of Science* **31** (1964) 319–327.

Hesse, M. B., 'Consilience of Inductions', in *The Problem of Inductive Logic* (ed. by I. Lakatos), pp. 232–246, North-Holland, Amsterdam, 1968.

Hesse, M. B., 'Ramifications of "Grue"', *British Journal for the Philosophy of Science* **20** (1969) 13–25. (Referred to as (1969a).)

Hesse, M. B., 'Confirmation of Laws', in *Philosophy Science, and Method* (ed. by S. Morgenbesser, P. Suppes, and M. White), pp. 74–91, St. Martin's Press, New York 1969. (Referred to as (1969b).)

Hesse, M. B., 'Theories and the Transitivity of Confirmation', *Philosophy of Science* **37** (1970) 50–63. (Referred to as (1970a).)

Hesse, M. B., 'An Inductive Logic of Theories', in *Analyses of Theories and Methods of Physics and Psychology, Minnesota Studies in the Philosophy of Science*, vol. IV (ed. by M. Radner and S. Winokur), pp. 164–180, University of Minnesota Press, Minneapolis 1970. (Referred to as (1970b).)

Hilpinen, R., 'On Inductive Generalization in Monadic First-Order Logic with Identity', in *Aspects of Inductive Logic* (ed. by K. J. Hintikka and P. Suppes), pp. 133–154, North-Holland, Amsterdam 1966.

Hilpinen, R., *Rules of Acceptance and Inductive Logic (Acta Philosophica Fennica* **21**), North-Holland, Amsterdam, 1968.

Hilpinen, R., 'On the Information Provided by Observations', in *Information and Inference* (ed. by K. J. Hintikka and P. Suppes), pp. 97–122, D. Reidel, Dordrecht, 1970.

Hilpinen, R., 'Relational Hypotheses and Inductive Inference', *Synthese* **23** (1971) 266–286.

Hilpinen, R., 'Decision-Theoretic Approaches to Rules of Acceptance', in *Contemporary Philosophy in Scandinavia* (ed. by R. E. Olson and A. M. Paul), pp. 147–168, The Johns Hopkins Press, Baltimore, 1972.

Hilpinen, R. and Hintikka, K. J., 'Rules of Acceptance, Indices of Lawlikeness, and Singular Inductive Inference: Reply to a Critical Discussion', *Philosophy of Science* **38** (1971) 303–307.

Hintikka, K. J., 'Towards a Theory of Inductive Generalization', in *Proceedings of the 1964 International Congress for Logic, Methodology, and Philosophy of Science* (ed. by Y. Bar-Hillel), pp. 274–288, North-Holland, Amsterdam 1965. (Referred to as (1965a).)

Hintikka, K. J., 'On a Combined System of Inductive Logic', in *Studia Logico-Mathematica et Philosophica in Honorem Rolf Nevanlinna, Acta Philosophica Fennica* **18** (1965) 21–30. (Referred to as (1965b).)

Hintikka, K. J., 'A Two-Dimensional Continuum of Inductive Methods', in *Aspects of Inductive Logic* (ed. by K. J. Hintikka and P. Suppes), pp. 113–132, North-Holland Amsterdam, 1966.

Hintikka, K. J., 'Induction by Enumeration and Induction by Elimination', in *The Problem of Inductive Logic* (ed. by I. Lakatos), pp. 191–216, North-Holland, Amsterdam 1968. (Referred to as (1968a).)

Hintikka, K. J., 'The Varieties of Information and Scientific Explanation', in *Logic, Methodology, and Philosophy of Science III, Proceedings of the 1967 International Congress* (ed. by B. van Rootselaar and J. F. Staal), pp. 151–171, North-Holland, Amsterdam, 1968. (Referred to as (1968b).)

Hintikka, K. J., 'Inductive Independence and the Paradoxes of Confirmation', in

Essays in Honor of Carl G. Hempel: A Tribute on the Occasion of his Sixty-Fifth Birthday (ed. by N. Rescher *et al.*), pp. 24–46, D. Reidel, Dordrecht 1969. (Referred to as 1969a).)

Hintikka, K. J., 'Statistics, Induction, and Lawlikeness: Comments on Dr. Vetter's Paper', *Synthese* **20** (1969) 72–85. (Referred to as (1969b).)

Hintikka, K. J., 'Tieteen metodi analyyttisenä toimituksena', in *Tieto on valtaa ja muita aatehistoriallisia esseitä*, pp. 272–293, Werner Söderström, Helsinki, 1969. (Referred to as (1969c).)

Hintikka, K. J., 'On Semantic Information', in *Information and Inference* (ed. by K. J. Hintikka and P. Suppes), pp. 3–27, D. Reidel, Dordrecht, 1970.

Hintikka, K. J., 'Inductive Generalization and its Problems: A Comment on Kronthaler's Comment', *Theory and Decision* **1** (1971) 393–398. (Referred to as (1971a).)

Hintikka, K. J., 'Unknown Probabilities, Bayesianism, and de Finetti's Representation Theorem', in *Boston Studies in the Philosophy of Science*, vol. 8 (ed. by R. Buck and R. S. Cohen), pp. 325–341, D. Reidel, Dordrecht, 1971. (Referred to as (1971b).)

Hintikka, K. J., 'On the Ingredients of an Aristotelian Science', *Noûs* **6** (1972) 55–69.

Hintikka. K. J., 'On the Different Ingredients of an Empirical Theory', *Logic, Methodology, and Philosophy of Science IV, Proceedings of the 1971 International Congress* (ed. by P. Suppes, L. Henkin, A. Joja, and Gr. Moisil), North-Holland, Amsterdam, 1973.

Hintikka, K. J. and Hilpinen, R., 'Knowledge, Acceptance, and Inductive Logic', in *Aspects of Inductive Logic* (ed. by K. J. Hintikka and P. Suppes), pp. 1–20, North-Holland, Amsterdam, 1966.

Hintikka, K. J. and Niiniluoto, I., 'On Theoretical Terms and Their Ramsey-Elimination: An Essay in the Logic of Science' (in Russian), *Filosofskiye Nauki* No. 1 (1973) 49–61.

Hintikka, K. J. and Pietarinen, J., 'Semantic Information and Inductive Logic', in *Aspects of Inductive Logic* (ed. by K. J. Hintikka and P. Suppes), pp. 96–112, North-Holland, Amsterdam, 1966.

Hintikka, K. J. and Tuomela, R., 'Towards a General Theory of Auxiliary Concepts and Definability in First-Order Theories', in *Information and Inference* (ed. by K. J. Hintikka and P. Suppes), pp. 92–124, D. Reidel, Dordrecht, 1970.

Hooker, C. A., 'Craigian Transcriptionism', *American Philosophical Quarterly* **5** (1968) 152–163.

Jeffrey, R. C., 'Valuation and Acceptance of Scientific Hypotheses', *Philosophy of Science* **23** (1956) 237–246.

Jeffrey, R. C., *The Logic of Decision*, McGraw-Hill, New York, 1965.

Jeffrey, R. C., 'Statistical Explanation vs. Statistical Inference', in *Essays in Honor of Carl G. Hempel* (ed. by N. Rescher *et al.*), pp. 104–113, D. Reidel, Dordrecht, 1969.

Jeffrey, R. C., 'Remarks on Explanatory Power', in *Boston Studies in the Philosophy of Science*, vol. VIII (ed. by R. C. Buck and R. S. Cohen), pp. 40–46, D. Reidel, Dordrecht, 1971.

Jevons, W. St., *The Principles of Science*, 2nd Edition, London, 1877.

Kaila, E., *Inhimillinen Tieto*, Otava, Helsinki, 1939.

Kemeny, J. G., 'The Use of Simplicity in Induction', *Philosophical Review* **62** (1953) 391–408.

Kemeny, J. G. and Oppenheim, P., 'On Reduction', *Philosophical Studies* **7** (1956) 6–19.

Kneale, W., *Probability and Induction*, Oxford University Press, Oxford, 1949.

Kuhn, T. S., *The Structure of Scientific Revolutions*, University of Chicago Press, Chicago, 1962. (2nd Edition, 1969)

Kyburg, H. E., 'The Rule of Detachment in Inductive Logic', in *The Problem of Inductive Logic* (ed. by I. Lakatos), pp. 98–119, North-Holland, Amsterdam, 1968.

Kyburg, H. E., *Probability and Inductive Logic*, The Macmillan Company, Collier-Macmillan Ltd, London, 1970.

Kyburg, H. E., Review of Cohen (1970), *Journal of Philosophy* **69** (1972) 106–114.

Lakatos, I., 'Changes in the Problem of Inductive Logic', in *The Problem of Inductive Logic* (ed. by I. Lakatos), pp. 315–417, North-Holland, Amsterdam, 1968.

Lehrer, K., 'Descriptive Completeness and Inductive Methods', *Journal of Symbolic Logic* **28** (1963) 157–160.

Lehrer, K., 'Theoretical Terms and Inductive Inference', *American Philosophical Quarterly*, Monograph Series, No. 3 (1969) 30–41.

Lehrer, K., 'Induction, Reason, and Consistency', *British Journal for the Philosophy of Science* **21** (1970) 103–114.

Lehrer, K., 'Induction and Conceptual Change', *Synthese* **23** (1971) 206–225.

Levi, I., *Gambling with Truth: An Essay on Induction and the Aims of Science*, Alfred A. Knopf, New York, 1967. (Referred to as (1967a).)

Levi, I., 'Information and Inference', *Synthese* **17** (1967) 369–391. (Referred to as (1967b).)

Levi, I., 'Confirmation, Linguistic Invariance, and Conceptual Innovation', *Synthese* **20** (1969) 48–55.

Levi, I., 'Probability and Evidence', in *Induction, Acceptance and Rational Belief* (ed. by M. Swain), pp. 134–156, D. Reidel, Dordrecht, 1970.

Mackie, J. L., 'A Simple Model of Consilience', in *The Problem of Inductive Logic* (ed. by I. Lakatos), pp. 250–253, North-Holland, Amsterdam, 1968.

Mackie, J. L., 'The Relevance Criterion of Confirmation', *British Journal for the Philosophy of Science* **20** (1969) 27–40.

Maxwell, G., 'Structural Realism and the Meaning of Theoretical Terms', in *Minnesota Studies in the Philosophy of Science*, vol. IV, *Analyses of Theories and Methods of Physics and Psychology* (ed. by M. Radner and S. Winokur), pp. 181–192, University of Minnesota Press, Minneapolis, 1970.

Maxwell, N., 'A Critique of Popper's View of Scientific Method', *Philosophy of Science* **39** (1972) 131–152.

Michalos, A. C., *The Popper-Carnap Controversy*, Martinus Nijhoff, The Hague, 1971.

Mill, J. S., *A System of Logic*, London 1843. (Quoted from *John Stuart Mill's Philosophy of Scientific Method*, ed. by E. Nagel, Hafner Publ. Co., New York, 1950.)

Nagel, E., *The Structure of Science*, Hartcourt, Brace and World, London, 1961.

Nagel, E., 'Carnap's Theory of Induction', in *The Philosophy of Rudolf Carnap* (ed. by P. A. Schilpp), Open Court, La Salle, 1963.

Nicod, J., *Foundations of Geometry and Induction*, Kegan Paul, Trench, Trubner & Co., London, 1930.

Niiniluoto, I., 'Can We Accept Lehrer's Inductive Rule?', *Ajatus* **33** (1971) 254–265.

Niiniluoto, I., 'Inductive Systematization: Definition and a Critical Survey', *Synthese* **25** (1972) 25–81. (Referred to as (1973a).)

Niiniluoto, I., 'Inductive Systematization and Empirically Trivial Theories', in *Logic, Language and Probability, A Selection of Papers Contributed to Sections 4, 6, and 11 of the IVth International Congress of Logic, Methodology and Philosophy of Science,*

Bucharest 1971 (ed. by R. J. Bogdan and I. Niiniluoto), pp. 108–114, D. Reidel, Dordrecht, 1973. (Referred to as (1973b).)

Niiniluoto, I., Review of Michalos (1971), *Synthese* 25 (1973) 417–436. (Referred to as (1973c).)

Pietarinen, J., 'Tools for Evaluating Scientific Systematizations', in *Information and Inference* (ed. by K. J. Hintikka and P. Suppes), pp. 123–147, D. Reidel, Dordrecht, 1970.

Pietarinen, J., *Lawlikeness, Analogy, and Inductive Logic* (*Acta Philosophica Fennica* 26), North-Holland, Amsterdam, 1972.

Pietarinen, J. and Tuomela, R., 'On Measures of the Explanatory Power of Scientific Theories', *Akten des 14. Internationalen Kongresses für Philosophie, Wien 2.-9. September 1968*, vol. 3, pp. 241–247, Universität von Wien, Verlag Herder, Wien, 1968.

Popper, K. R., *The Logic of Scientific Discovery*, Hutchinson & Co., London, 1959. (Quoted from the revised edition, 1968.)

Popper, K. R., *Conjectures and Refutations*, Routledge and Kegan Paul, London, 1963. (Quoted from the third revised edition, 1969.)

Popper, K. R., *Objective Knowledge, An Evolutionary Approach*, Oxford University Press, Oxford, 1972.

Putnam, H., 'Degrees of Confirmation and Inductive Logic', in *The Philosophy of Rudolf Carnap* (ed. by P. A. Schilpp), pp. 761–784, Open Court, La Salle, 1963.

Putnam, H., 'Craig's Theorem', *Journal of Philosophy* 62 (1965) 251–260.

Quine, W. V. O., 'On Simple Theories of a Complex World', *Synthese* 15 (1963) 103–106.

Ramsey, F. P., 'Theories', in *The Foundations of Mathematics and Other Logical Essays*, pp. 212–236, Littlefield, Adams & Co., Paterson, 1931.

Rescher, N., 'Lawfulness as Mind-Dependent', in *Essays in Honor of Carl G. Hempel: A Tribute on the Occasion of his Sixty-Fifth Birthday* (ed. by N. Rescher *et al.*), pp. 178–197, D. Reidel, Dordrecht, 1969.

Rosenkrantz, R. D., 'On Explanation', *Synthese* 20 (1969) 335–370.

Rosenkrantz, R. D., 'Experimentation as Communication with Nature', in *Information and Inference* (ed. by K. J. Hintikka and P. Suppes), pp. 58–93, D. Reidel, Dordrecht, 1970.

Rozeboom, W., 'The Art of Metascience', in *Toward Unification in Psychology* (ed. by J. R. Royce), pp. 53–163, University of Toronto Press, Toronto, 1970.

Rudner, R., 'An Introduction to Simplicity', *Philosophy of Science* 28 (1961) 109–119.

Salmon, W. C., 'Vindication of Induction', in *Current Issues in the Philosophy of Science* (ed. by H. Feigl and G. Maxwell), pp. 245–257, Holt, Rinehart and Winston, New York, 1961. (Referred to as (1961a).)

Salmon, W. C., 'Rejoinder to Barker', in *Current Issues in the Philosophy of Science* (ed. by H. Feigl and G. Maxwell), pp. 260–262, Holt, Rinehart and Winston, New York, 1961. (Referred to as (1961b).)

Salmon, W. C., 'On Vindicating Induction', in *Induction: Some Current Issues* (ed. by H. E. Kyburg and E. Nagel), pp. 27–41, Wesleyan University Press, Middletown, 1963. (Referred to as (1963a).)

Salmon, W. C., 'Reply to Black', in *Induction: Some Current Issues* (ed. by H. E. Kyburg and E. Nagel), pp. 49–54, Wesleyan University Press, Middletown, 1963. (Referred to as (1963b).)

Salmon, W. C., 'Consistency, Transitivity, and Inductive Support', *Ratio* 7 (1965) 164–169.

Salmon, W. C., *The Foundations of Scientific Inference*, University of Pittsburgh Press, Pittsburgh, 1966.

Salmon, W. C., 'Carnap's Inductive Logic', *Journal of Philosophy* **64** (1967) 725–739.

Salmon, W. C., 'The Justification of Inductive Rules of Inference', in *The Problem of Inductive Logic* (ed. by I. Lakatos), pp. 24–43, North-Holland, Amsterdam, 1968.

Salmon, W. C., 'Statistical Explanation', in *Nature and Function of Scientific Theories* (ed. by R. G. Colodny), pp. 173–231, University of Pittsburgh Press, Pittsburgh, 1970, Reprinted in *Statistical Explanation and Statistical Relevance* (ed. by W. C. Salmon et al.), pp. 29–87, University of Pittsburgh Press, Pittsburgh, 1971. (Referred to as (1971a).)

Salmon, W. C., 'Explanation and Relevance: Comments on James G. Greeno's "Theoretical Entities in Statistical Explanation"', in *Boston Studies in the Philosophy of Science*, vol. VIII (ed. by R. C. Buck and R. S. Cohen), pp. 27–39, D. Reidel, Dordrecht, 1971. (Referred to as (1971b).)

Scheffler, I., *The Anatomy of Inquiry: Philosophical Studies in the Theory of Science*, Alfred A. Knopf, New York, 1963.

Scheffler, I., *Science and Subjectivity*, Bobbs-Merrill, New York, 1967.

Scheffler, I., 'Reflections on the Ramsey Method', *Journal of Philosophy* **65** (1968) 269–274.

Schlesinger, G., 'On Irrelevant Criteria of Confirmation', *British Journal for the Philosophy of Science* **21** (1970).

Sellars, W., *Science, Perception, and Reality*, Routledge and Kegan Paul, London, 1963.

Sellars, W., 'Scientific Realism or Irenic Instrumentalism', in *Boston Studies in the Philosophy of Science*, vol. II (ed. by R. S. Cohen and M. W. Wartofsky), Humanities Press, New York, 1965, (Reprinted in Sellars, W., *Philosophical Perspectives*, pp. 337–369, 1967).

Shannon, C. E. and Weaver, W., *The Mathematical Theory of Communication*, The University of Illinois Press, Urbana, 1949.

Sherwood, M. M., *The Logic of Explanation in Psychoanalysis*, Academic Press, New York, 1969.

Smokler, H., 'Conflicting Conceptions of Confirmation', *Journal of Philosophy* **65** (1968) 300–312.

Sneed, J. D., *The Logical Structure of Mathematical Physics*, D. Reidel, Dordrecht, 1971.

Spector, M., 'Theory and Observation I–II', *British Journal for the Philosophy of Science* **7** (1966) 1–20, 89–104.

Stalnaker, R. C. and Thomason, R. H., 'A Semantic Analysis of Conditional Logic', *Theoria* **34** (1970) 23–42.

Stegmüller, W., *Wissenschaftliche Erklärung und Begründung, Probleme und Resultate der Wissenschafstheorie und Analytischen Philosophie*, Band I, Springer-Verlag, Berlin, Heidelberg & New York, 1969.

Stegmüller, W., *Theorie und Erfahrung, Probleme und Resultate der Wissenschaftstheorie und Analytischen Philosophie*, Band II, Springer-Verlag, Berlin, Heidelberg & New York, 1970.

Stegmüller, W., 'Das Problem der Induktion: Humes Herausforderung und moderne Antworten', in *Neue Aspekte der Wissenschaftstheorie* (ed. by H. Lenk), Vieweg, Braunschweig, 1971.

Suppes, P., 'Concept Formation and Bayesian Decisions', in *Aspects of Inductive Logic* (ed. by K. J. Hintikka and P. Suppes), pp. 21–48, North-Holland, Amsterdam, 1966.

Törnebohm, H., 'Two Measures of Evidential Strength', in *Aspects of Inductive Logic* (ed. by K. J. Hintikka and P. Suppes), pp. 81–95, North-Holland, Amsterdam, 1966.

Tuomela, R., 'Inductive Generalization in an Ordered Universe', in *Aspects of Inductive Logic* (ed. by K. J. Hintikka and P. Suppes), pp. 155–174, North-Holland, Amsterdam, 1966.

Tuomela, R., 'The Role of Theoretical Concepts in Neobehavioristic Theories', *Reports from the Institute of Philosophy, University of Helsinki*, No. 1, 1971.

Tuomela, R., 'Deductive Explanation of Scientific Laws', *Journal of Philosophical Logic* 1 (1972) 369–392.

Tuomela, R., *Theoretical Concepts*, LEP Vol. 10, Springer-Verlag, Wien, 1973.

Uchii, S., 'Inductive Logic with Causal Modalities: A Probabilistic Approach', *Philosophy of Science* 39 (1972) 162–178.

Watkins, J. W. N., 'Non-inductive Corroboration', in *The Problem of Inductive Logic* (ed. by I. Lakatos), pp. 61–66, North-Holland, Amsterdam, 1968.

Whewell, W., *The History of Scientific Ideas*, London, 1847.

Whewell, W., *Novum Organon Renovatum*, London, 1858.

Wright, G. H. von, *A Treatise on Induction and Probability*, Routledge and Kegan Paul, London, 1951.

Wright, G. H. von, *The Logical Problem of Induction*, 2nd revised edition, Basil Blackwell, Oxford, 1957.

Wright, G. H. von, *Explanation and Understanding*, Cornell University Press, Cornell, 1971.

INDEX OF NAMES

INDEX OF SUBJECTS

SYNTHESE LIBRARY

Monographs on Epistemology, Logic, Methodology,

Philosophy of Science, Sociology of Science and of Knowledge, and on the

Mathematical Methods of Social and Behavioral Sciences

Editors:

DONALD DAVIDSON (The Rockefeller University and Princeton University)

JAAKKO HINTIKKA (Academy of Finland and Stanford University)

GABRIËL NUCHELMANS (University of Leyden)

WESLEY C. SALMON (Indiana University)

MARIO BUNGE (ed.), *Exact Philosophy. Problems, Tools, and Goals.* 1973, X + 214 pp.

ROBERT S. COHEN and MARX W. WARTOFSKY (eds.), *Boston Studies in the Philosophy of Science.* Volume IX: *A. A. Zinov'ev: Foundations of the Logical Theory of Scientific Knowledge (Complex Logic).* Revised and Enlarged English Edition with an Appendix by G. A. Smirnov, E. A. Sidorenka, A. M. Fedina, and L. A. Bobrova. 1973, XXII + 301 pp. Also available as a paperback.

K. J. J. HINTIKKA, J. M. E. MORAVCSIK, and P. SUPPES (eds.), *Approaches to Natural Language. Proceedings of the 1970 Stanford Workshop on Grammar and Semantics.* 1973, VIII + 526 pp. Also available as a paperback.

WILLARD C. HUMPHREYS, JR. (ed.), *Norwood Russell Hanson: Constellations and Conjectures.* 1973, X + 282 pp.

MARIO BUNGE, *Method, Model and Matter.* 1973, VII + 196 pp.

MARIO BUNGE, *Philosophy of Physics.* 1973, IX + 248 pp.

LADISLAV TONDL, *Boston Studies in the Philosophy of Science.* Volume X: *Scientific Procedures.* 1973. XIII + 268 pp. Also available as a paperback.

SÖREN STENLUND, *Combinators, λ-Terms and Proof Theory.* 1972, 184 pp.

DONALD DAVIDSON and GILBERT HARMAN (eds.), *Semantics of Natural Language.* 1972, X + 769 pp. Also available as a paperback.

MARTIN STRAUSS, *Modern Physics and Its Philosophy. Selected Papers in the Logic, History, and Philosophy of Science.* 1972, X + 297 pp.

‡STEPHEN TOULMIN and HARRY WOOLF (eds.), *Norwood Russell Hanson: What I Do Not Believe, and Other Essays,* 1971, XII + 390 pp.

‡ROBERT S. COHEN and MARX W. WARTOFSKY (eds.), *Boston Studies in the Philosophy of Science.* Volume VIII: *PSA 1970. In Memory of Rudolf Carnap* (ed. by Roger C. Buck and Robert S. Cohen). 1971, LXVI + 615 pp. Also available as a paperback.

‡YEHOSUA BAR-HILLEL (ed.), *Pragmatics of Natural Languages* 1971, VII + 231 pp.

‡ROBERT S. COHEN and MARX W. WARTOFSKY (eds.), *Boston Studies in the Philosophy of Science.* Volume VII: *Milič Čapek: Bergson and Modern Physics.* 1971, XV + 414 pp.

‡CARL R. KORDIG, *The Justification of Scientific Change.* 1971, XIV + 119 pp.

‡JOSEPH D. SNEED, *The Logical Structure of Mathematical Physics.* 1971, XV + 311 pp.

‡JEAN-LOUIS KRIVINE, *Introduction to Axiomatic Set Theory.* 1971, VII + 98 pp.

‡RISTO HILPINEN (ed.), *Deontic Logic: Introductory and Systematic Readings.* 1971, VII + 182 pp.

‡EVERT W. BETH, *Aspects of Modern Logic.* 1970, XI + 176 pp.

‡PAUL WEINGARTNER and GERHARD ZECHA, (eds.), *Induction, Physics, and Ethics, Proceedings and Discussions of the 1968 Salzburg Colloquium in the Philosophy of Science.* 1970, X + 382 pp.

‡ROLF A. EBERLE, *Nominalistic Systems.* 1970, IX + 217 pp.

‡JAAKKO HINTIKKA and PATRICK SUPPES, *Information and Inference.* 1970, X + 336 pp.

‡KAREL LAMBERT, *Philosophical Problems in Logic. Some Recent Developments.* 1970, VII + 176 pp.

‡P. V. TAVANEC (ed.), *Problems of the Logic of Scientific Knowledge.* 1969, XII + 429 pp.

‡ROBERT S. COHEN and RAYMOND J. SEEGER (eds.), *Boston Studies in the Philosophy of Science.* Volume VI: *Ernst Mach: Physicist and Philosopher.* 1970, VIII + 295 pp.

‡MARSHALL SWAIN (ed.), *Induction, Acceptance and Rational Belief.* 1970, VII + 232 pp.

‡NICHOLAS RESCHER *et al.* (eds.), *Essays in Honor of Carl G. Hempel. A Tribute on the Occasion of his Sixty-Fifth Birthday.* 1969, VII + 272 pp.

‡PATRICK SUPPES, *Studies in the Methodology and Foundations of Science. Selected Papers from 1911 to 1961.* 1969, XII + 473 pp.

‡JAAKKO HINTIKKA, *Models for Modalities. Selected Essays.* 1969, IX + 220 pp.

‡D. DAVIDSON and J. HINTIKKA (eds.), *Words and Objections: Essays on the Work of W. V. Quine.* 1969, VIII + 366 pp.

‡J. W. DAVIS, D. J. HOCKNEY and W. K. WILSON (eds.), *Philosophical Logic.* 1969, VIII + 277 pp.

‡ROBERT S. COHEN and MARX W. WARTOFSKY (eds.), *Boston Studies in the Philosophy of Science*, Volume V: *Proceedings of the Boston Colloquium for the Philosophy of Science 1966/1968*, VIII + 482 pp.

‡ROB ERTS. COHEN and MARX W. WARTOFSKY (eds.), *Boston Studies in the Philosophy of Science.* Volume IV: *Proceedings of the Boston Colloquium for the Philosophy of Science 1966/1968.* 1969, VIII + 537 pp.

‡NICHOLAS RESCHER, *Topics in Philosophical Logic.* 1968, XIV + 347 pp.

‡GÜNTHER PATZIG, *Aristotle's Theory of the Syllogism. A Logical-Philological Study of Book A of the Prior Analytics.* 1968, XVII + 215 pp.

‡C. D. BROAD, *Induction, Probability, and Causation. Selected Papers.* 1968, XI + 296 pp.

‡ROBERT S. COHEN and MARX W. WARTOFSKY (eds.), *Boston Studies in the Philosophy of Science.* Volume III: *Proceedings of the Boston Colloquium for the Philosophy of Science 1964/1966.* 1967, XLIX + 489 pp.

‡GUIDO KÜNG, *Ontology and the Logistic Analysis of Language. An Enquiry into the Contemporary Views on Universals.* 1967, XI + 210 pp.

*EVERT W. BETH and JEAN PIAGET, *Mathematical Epistemology and Psychology*. 1966, XXII + 326 pp.

*EVERT W. BETH, *Mathematical Thought. An Introduction to the Philosophy of Mathematics*. 1965, XII + 208 pp.

‡PAUL LORENZEN, *Formal Logic*. 1965, VIII + 123 pp.

‡GEORGES GURVITCH, *The Spectrum of Social Time*. 1964, XXVI + 152 pp.

‡A. A. ZINOV'EV, *Philosophical Problems of Many-Valued Logic*. 1963, XIV + 155 pp.

‡MARX W. WARTOFSKY (ed.), *Boston Studies in the Philosophy of Science*. Volume I: *Proceedings of the Boston Colloquium for the Philosophy of Science, 1961–1962*. 1963, VIII + 212 pp.

‡B. H. KAZEMIER and D. VUYSJE (eds.), *Logic and Language. Studies dedicated to Professor Rudolf Carnap on the Occasion of his Seventieth Birthday*. 1962, VI + 256 pp.

*EVERT W. BETH, *Formal Methods. An Introduction to Symbolic Logic and to the Study of Effective Operations in Arithmetic and Logic*. 1962, XIV + 170 pp.

*HANS FREUDENTHAL (ed.), *The Concept and the Role of the Model in Mathematics and Natural and Social Sciences. Proceedings of a Colloquium held at Utrecht, The Netherlands, January 1960*. 1961, VI + 194 pp.

‡P. L. GUIRAUD, *Problèmes et méthodes de la statistique linguistique*. 1960, VI + 146 pp.

*J. M. BOCHEŃSKI, *A Precis of Mathematical Logic*. 1959, X + 100 pp.

SYNTHESE HISTORICAL LIBRARY

Texts and Studies
in the History of Logic and Philosophy

Editors:

N. KRETZMANN (Cornell University)
G. NUCHELMANS (University of Leyden)
L. M. DE RIJK (University of Leyden)

LEWIS WHITE BECK (ed.), *Proceedings of the Third International Kant Congress.* 1972, XI + 718 pp.

‡KARL WOLF and PAUL WEINGARTNER (eds.), *Ernst Mally: Logische Schriften.* 1971, X + 340 pp.

‡LEROY E. LOEMKER (ed.), *Gottfried Wilhelm Leibnitz: Philosophical Papers and Letters.* A Selection Translated and Edited, with an Introduction. 1969, XII + 736 pp.

‡M. T. BEONIO-BROCCHIERI FUMAGALLI, *The Logic of Abelard.* Translated from the Italian. 1969, IX + 101 pp.

Sole Distributors in the U.S.A. and Canada:

*GORDON & BREACH, INC., 440 Park Avenue South, New York, N.Y. 10016
‡HUMANITIES PRESS, INC., 303 Park Avenue South, New York, N.Y. 10010